Activity Guide

for Wiley Geography Textbooks and Microsoft® Encarta® Virtual Globe

Second Edition

Margaret M. Gripshover
Marshall University

Thomas L. Bell
The University of Tennessee, Knoxville

JOHN WILEY & SONS, INC.
New York • Chichester • Weinheim • Brisbane • Singapore • Toronto

To order books or for customer service call 1-800-CALL-WILEY (225-5945).

ISBN 0-471-37795-3

Printed in the United States of America

10 9 8 7 6 5 4 3 2 1

Printed and bound by Victor Graphics, Inc.

Dedication

This edition is dedicated to our mothers who instilled in us
a great curiosity about the world around us.

Alice Joyce Brennen
and
Lilah Mergy Bell

Acknowledgements

We wish to express our sincere appreciation to our editor Nanette Kauffman, and assistant editor, Alicia Solis, for their cheerfulness, patience, and goodwill throughout the entire process of bringing this volume from concept to completion. It was really Nanette's vision that guided the process throughout and we hope we have been faithful to her vision of showcasing the incredible range of activities that one can accomplish using Microsoft's® *Encarta® Virtual Globe '99* CD-ROMs. We have become believers in the software product and are continually amazed by its flexibility and comprehensiveness. The decision to package the textual material with the CD-ROMs and *Wiley Activity Guide* was a good one. If the proof of the pudding is in the tasting, we hope the proof of the utility of this software is its continued use as a valuable reference tool throughout the student's college career and beyond.

We also want to thank the geography departments at Marshall University and The University of Tennessee, Knoxville for providing us with the space and resources to complete the project. Kyle Rector and Terry Gilhula, Ph.D. candidate and freshly minted Ph.D. in geography at the University of Tennessee respectively, deserve many thanks for making sure that the computers used in this project were installed properly and up-to-date.

Table of Contents

Introduction

John Wiley and Sons, Inc., has entered into an exciting educational agreement with Microsoft® to package the latest version of Microsoft's® *Encarta® Virtual Globe '99* CD-ROMs with many popular Wiley undergraduate textbooks in geography. *Encarta® Virtual Globe '99* is an interactive electronic atlas that combines traditional thematic maps with video clips, slides, sound clips, data sets and so much more. This *Wiley Activity Guide for the Use of Encarta's® Virtual Globe '99 CD-ROMs* is included with the software. We have prepared this *Activity Guide* to provide faculty and students with a learning tool that connects the software with the material in the textbooks. Exercises or activities are included for each chapter (or section) of nine best-selling Wiley textbooks.[1] These books include three introductory physical geography textbooks with approximately the same content sequencing (Strahler and Strahler 1997; Strahler and Strahler 1998; de Blij and Muller 1997), a regional geography textbook (de Blij and Muller 2000) and two human geography texts (de Blij and Murphy 1999; Kuby, *et al.*, 1998). Activities using the *Encarta® Virtual Globe '99* CD-ROMs were also developed for more specialized texts in economic geography (Wheeler, *et al.* 1998) and the regional geography of Latin America (Blouet and Blouet, 1996) and the United States and Canada (Birdsall, *et al.*, 1999).

We hope to engage students in a positive interactive experience rather than using *Encarta® Virtual Globe '99* as a place-name memorization tool. We will have succeeded in our mission if students feel compelled to incorporate the material from the *Virtual Globe '99* CD-ROMs in other contexts beyond the particular course for which they first purchased them. And, for those students who are somewhat techno- or cyber-phobic, the *Wiley Activity Guide, Second Edition,* will ease them into displaying data in both mapped and tabular form and drawing the appropriate conclusions from those data.

Students may also take virtual fieldtrips via a wonderful set of slides, videos and sound clips that can be accessed for various countries or themes. Some of the activities employ the use of the *measuring tool* that can be used to determine the straight line distance between any pair of nodes on a map produced at any given scale. Within the context of a given activity, students might be asked to go to the *Web Links* associated with particular themes or areas. When connected by modem or Ethernet connection, the *Encarta® Virtual Globe '99* CD-ROMs have a built-in browser (Microsoft's® *Internet Explorer*) that catalogues relevant *Web Links*. Should a student or instructor prefer to access web sites using other browsers (e.g., *Netscape*) that is perfectly acceptable as well and quite easily accomplished when the programs are set-up and initialized.

Instructors may add relevant links that are not included among the pre-selected web sites to the list. For example, in an activity that focuses on the "Great Migration" of African Americans from the rural South to the urban North, a web link is included in the *Wiley Activity Guidw, Second Edition*, to a Library of Congress site. That site focuses specifically on the impact of the African American migration stream on the city of Chicago. The web site references a current exhibition that highlights the black experience.

---

[1] Due to space limitations, only five chapters of Birdsall, *et al.*'s excellent text, *Regional Landscapes of The United States and Canada, Fifth Edition* are included.

Anyone who has tried to use the Internet as a teaching tool has probably encountered the problem of ephemeral web sites. The sites that *Encarta® Virtual Globe '99* has selected will probably be no exception. But, many of the *Encarta®* links that are pre-loaded into the CD-ROMs are official sources or reputable and well-established commercial enterprises and are less likely to suddenly disappear.

What do the activities and exercises in the *Wiley Activity Guide, Second Edition* consist of exactly? Each is relatively short (about 2-3 pages) with a focused set of questions students can answer after they explore some highlighted aspect of *Encarta® Virtual Globe '99*. They may, for example, be asked to compare Bantu music with that of the Hausa using the *Sights and Sounds* feature. Or, they may be taken to a *web link* that describes the descent into Mammoth Cave by hearty souls in the early 19th century when the cave was still in private hands and compare the adventure to organized tours through the well-lit and paved trails of the National Park today. Likewise, they may be asked to assess the mapped correlation between infant mortality rates and total fertility rate at the world scale using both maps and data arrayed in tabular form. They may see for themselves the mistake that Dutch planners made by hewing too closely to the spacing requirements of Christaller's classic central place model when settlements were located on reclaimed polderlands. They might realize that rock(s) in Australia could refer to the sacred Ayres (Uluru) Rock of the Aborigines, the place of first settlement in Sydney, or the many internationally famous rock bands that have gotten their start in that country. At the end of each activity is a list of keywords that define its regional and thematic content.

The Importance of the Index

These keywords are cross-referenced in an index so that instructors are not limited to assigning activities from one textbook only. Suppose an instructor is interested in the cross-cultural comparison of urban spatial structure. He or she might have students examine the presentation of the Griffin-Ford model for the Latin American city in Blouet and Blouet's Latin American textbook as well as the coverage of the restructured metropolis in Wheeler, *et al.* regarding the North American city.

System Requirements and a Few Suggestions

Microsoft's® *Encarta® Virtual Globe '99* CD-ROMs are designed to run in a Microsoft® Windows environment (3.1, NT, or Windows '95 and later) and must have at least 8 MBs of RAM available although it works most efficiently with a full 32 MBs of RAM available.

Microsoft's® *Encarta® Virtual Globe '99* is at its most useful when speakers and a sound card are part of the computer's configuration. It can be used without sound (a warning message to that effect is posted) but there are many musical clips and other sounds that add to the overall enjoyment of the software.

The software **must be installed** the first time Microsoft's® *Encarta® Virtual Globe '99* is inserted into a machine that has not previously played the software. This takes only about a minute and even those who have never installed a piece of software before will find this whole process to be quite painless and user-friendly. Every school and university system will be configured differently, so it would behoove instructors to make sure the software is actually installed on all computers, to avoid any unnecessary glitches or delays that might frustrate students. The software is very logically constructed and "forgiving". By the latter term we mean that there are several routes that might be taken to achieve a desired end. It is advisable to devote one period just for students to "play" with the features of the CD-ROM in a group setting (e.g., a computer laboratory). There is a synergistic effect that will reverberate throughout the room and excite even the seemingly uninterested student in a way that other media cannot.

Features that Students will Undoubtedly Discover on Their Own

There are features of the CD-ROM that are not included in our activities. For example, there is a built-in "Name that Place" geography quiz that may be played individually or perhaps in competition. There are five levels of expertise in this game and the twenty questions that comprise a round are randomly generated. For those who enjoy *Jeopardy!* or *Trivial Pursuit*, this game will be addictive.

Likewise, there are several *world flights* included in the software that allow a student "pilot" to fly over a particular corridor and view the landscape below. Only a few of these flight paths have been used in the *Wiley Activity Guide, Second Edition* activities, but instructors may wish to incorporate more as they see fit. The instrument controls may be adjusted to fly lower or higher, more slowly or more rapidly, or in a changed direction if desired. Students will probably discover the delights of this feature of *Encarta's®* *Virtual Globe '99* on their own. We hope your students will enjoy *Encarta's® Virtual Globe '99* CD-ROMs and that the accompanying *Wiley Activity Guide, Second Edition* that we have written, will save you valuable exercise preparation time. Have fun!

de Blij and Muller, *Geography: Realms, Regions, and Concepts*, 9e

Introduction: World Regional Geography

Back in the USSR: Georgia on My Mind

Most geographic realms can be sharply delimited such as those illustrated in Figure 1-1 in your text. Most students could identify North America, South America and Australia. You may even be able to discern the difference between East Asia and South Asia. But what of areas with less distinct boundaries such as the three transition zones on the perimeter of the North Africa/Southwest Asia realm? Where are these areas and why are they identified as "transition zones?" Two of these zones border the Russia realm and continue to be haunted by the legacy of decades of Soviet rule. Not only did the Soviets impose their governmental, social and economic influence over these regions, but they also attempted to dilute the indigenous populations culturally by resettling ethnic Russians within the territories.

To gain a better insight of the complexity of transition zones, take a closer look at the Trans-Caucasus. Using *Find*, create of map of the region by searching for Caucasus Mountains. Three former Soviet republics, Armenia, Azerbaijan, and Georgia, are now independent states. But there are conflicts among these states stemming from religious, ethnic, linguistic and cultural differences. One area of contention between Armenia and Azerbaijan is a territory called Nagorno-Karabakh. Click on *Geography*.

1. Based on the written material within *Geography*, describe in your own words the nature of this territorial dispute between Armenia and Azerbaijan.

2. For each of these three newly independent states, use *Facts and Figures* to find the proportion of the population that adheres to various religions. Write those proportions here.

3. Do you think that the fact that the vast majority of Armenians are Christian (i.e., Armenian Orthodox) and the vast majority of Azeris (residents of Azerbaijan) are Muslim could be one of the sources of irritation in their territorial dispute over Nogorno-Karabakh? Why or why not? Can you name the President of Georgia (hint: it is NOT Jimmy Carter)? Click on *Society*. Then click on *Government*.

4. What interesting linkages are there between the government of the former Soviet Union and the independent Republic of Georgia?

1

5. What position did the current President of Georgia hold during the Soviet period?
Using *Facts and Figures*, compare the economy of Georgia with those of Russia and Turkey.
We are assuming that Turkey is representative of many of the countries in the North
Africa/Southwest Asia realm.

6. In what ways is the economic structure of Georgia closer to that of Russia?

7. In what ways is it closer to that of Turkey?

Keywords: **transition zone, Russia, Transcaucasus, Georgia, Armenia, Azerbaijan**

Von Thünen on Whole Wheat? Contemporary European Agricultural Land Use

One of the classic location theories is that of the German economist Johann Heinrich von Thünen. Not only was von Thünen a brilliant scholar who contributed to the field of microeconomics with his notions about marginal returns to investment, but he also was a very successful gentleman farmer who owned a large estate called Tellow in northern Germany near the present day city of Rostock. The principles of agricultural land use he developed were based on many of his own observations of the workers on his estate and the amount of inputs (e.g., seed, fertilizer) they applied to the land to bring it into production.

Von Thünen's theory was presented in his book *The Isolated State* written in 1826 and fully translated into English in 1966. Any location theory is a product of its times and when von Thünen wrote about his land use model, transportation was limited overland to carts drawn by draft animals. Thus, the geographic range at which the theory was applicable was limited in size. Modern geographers have argued that von Thünen's basic principles--distance from market center being the main determinant of agricultural land uses and land use intensity diminishing with distance from the market place--are still true today, although at a scale of analysis not dreamed of by von Thünen himself. These models are sometimes referred to as macro-Thünen models.

One such macro-Thünen model is presented in your textbook (Figure 1-4, p. 53, in the 8th edition; Figure 1-6, p. 53, in the 9th edition). In the previous figure (Figure 1-3, p. 52, in the 8th edition; Figure 1-5, p. 52, in the 9th edition), the hypothetical land use pattern shown is predicted by purely economic principles. These principles were developed by von Thünen and assume no topographic barriers or place-to-place differences in soil fertility. Figure 1-4 (Figure 1-6, in the 9th edition) is based on the yields of eight leading crops and was produced by van Valkenburg and Held many decades ago. In the 1960s, studies of agricultural land uses in Europe were updated, but is the pattern shown in Figure 1-4 (Figure 1-6, in the 9th edition) still relevant in the 1990s and beyond?

One might argue that with the advent of the European Union, and the reduction of tariff barriers on agricultural products among the member states, that agriculture in Europe will begin to specialize to a much greater degree than has been the case in the past. Using the statistics available on Microsoft® *Encarta*® *Virtual Globe*, let's see if the land use patterns presented in Figure 1-4 (Figure 1-6, in the 9th edition)are still in evidence using four countries that fall wholly or mostly within the hypothesized agricultural zone.

Intensive farming and dairying may be represented by Denmark, the forest zone by Germany, field crops by Hungary and ranching and animal products by Spain. Using *Map Styles*, select *Statistical.* Under the *Features* section at the top of the page, choose *Statistics Center*. Then select the category *Agriculture* to examine spatial variation in the four representative crops or land uses discussed above. For intensive farming and dairying let's use *Livestock, Cattle 1997*. In Europe, most cattle serve the dual purpose of providing dairy products and meat. In a larger country that can afford the luxury of agricultural specialization like the United States, some breeds (e.g., Holstein, Jersey, Guernsey) have been bred for milk production almost exclusively and others (e.g., Angus, shorthorns) for meat production. At the bottom of the screen, select Statistical Table so that the world data may be displayed in the form of a bar chart. You will probably wish to sort the data in the table alphabetically and develop your own table of data for the four countries in question. At the bottom of the screen Microsoft® Encarta® allows two sorting methods—by rank and alphabetically by name of country. Select the latter--*Alphabetical* sorting. In order to gauge the relative importance of an agricultural activity relative to the population of the country, you will have to return to *Statistical Center* and choose the *Population* category. From there, scroll down to the statistic called *Population, total*. This figure will serve as the denominator for expressing the importance of the various agricultural and forest resource activities on a per capita basis.

1. Which of the four countries had the greatest number of cattle in 1997?

2. In which of the four countries are cattle most important to the economy based on a per capita basis? List the per capita figures on cattle in 1997 for the four countries of interest.

For the forest zone, we focus on two different statistics—within *Agriculture*, look at *Forest, area (thousand hectares)* and also examine *Sawn wood production, (cubic meters)*. Repeat the procedure above (i.e., calculate both absolute and relative values for each of the two measures).

3. Why do you think that there is a discrepancy between the proportion of total land in a country devoted to forest and woodland and the actual harvesting of those woodlands for lumber (e.g., sawn wood)?

For field crops let's choose *Wheat produced, metric tons* under *Agriculture*. Maize (what we call corn) does not grow well in Europe due to the lack of intensive solar insulation. But, Europe grows many short grain crops such as rye and tuber crops such as potatoes that have short growing seasons and can be planted in the poorer soils of northern Europe and the Scandinavian Peninsula.

4. Relatively speaking, where is wheat production most important to the agricultural systems of the four countries examined?

For the grazing of animals let's use *Livestock, sheep (numbers)* under *Agriculture* as sheep and goats can be pastured on poorer quality forage than can cattle and might more likely be found in agriculturally marginal areas.

5. Do the statistics support the contention that the zones shown in Figure 1-4 (Figure 1-6, in the 9th edition)still relevant in the 1990s and beyond? That is, does Denmark rank relatively higher in cattle production than, say, Hungary or Spain?

6. How about sheep production in Spain? Is it relatively more important to the economy of that country than say, Hungary or Denmark?

7. If you consider only the four countries of interest (i.e., if you examine their rank relative to each other and not to all the other countries of the world) is the model of van Valkenberg and Held (Figure 1-4 or Figure 1-6) still valid? Why or why not?

Keywords: **Europe, von Thünen, model, agricultural land use**

Chapter 2: Russia

Siberia: Many Rivers Run Through It

Siberia means many things, most of which are extremes to many people. For example, when winter weather turns frigid, meteorologists often liken the weather to Siberia. Imagine living in Noril'sk, the Siberian city whose major claim to fame is being the coldest city in the world with a mean annual temperature of 12 degrees Fahrenheit. For centuries, Russian governments have exiled political prisoners and convicted criminals to Siberia with the notion that escapes through thousands of miles of desolate territory was nearly impossible to survive. Basically, anything hard, vast, bitterly cold, or Doctor Zhivago-like, has been associated with Siberia.

PART I: Siberia, The Region

Siberia comprises a tremendous amount of territory in north-central Russia—over five million square miles of some of the most inhospitable land on the planet. Yet, Siberia is home to nearly 40 million residents. To locate Siberia using Virtual Globe 99, use the *Find* box and then click on "Siberia (region) Russia" in the *Places* window. Using the *Map Styles* option, select *Physical Features*. As you view the map, one of the most striking features is the three major rivers that drain the region. Before we navigate the rivers of Siberia, let's take a closer look at the region. Click on *Geography* and read a brief description of the region and answer the following questions.

1. What is the origin of the region's name and what does it mean in that language? Do you think that name applies to Siberia today---why or why not?

2. What is the main land transportation link in Siberia and how has this affected urbanization and economic development?

3. Of the over 150,000 rivers in Siberia, they all have one thing in common---what would that be?

For an up close and personal look at Siberia, click on *Sights and Sounds* and view the seven slides and listen to two ethnic folk music samples and answer these questions.

4. What are the ethnic influences associated with Sakha music? How does this differ from Shaman music?

5. Why do you think shamanist ceremonies are needed to help prevent young people from committing suicide---why would such a social crisis exist in Siberia?

6. What are some of the natural resources that are extracted from Siberia? How have these operations affected the bog environment in the Urmany region? What sort of environmental degradation has taken place in Siberia and do you think these problems will continue in the future---why or why not?

PART II: Siberia's Rivers

7. List the names of the three major rivers that drain most of the Siberia Region. (Hint: One has a very large tributary, the Irtysh.)

a.

b.

c.

8. How would you describe the direction of stream flow (i.e. where are the headwaters and the mouths of the rivers) and what are some problems that could result from this type of drainage pattern? For more background information on the physical geography of the region, you can read the *Geography* discussion for Siberia.

Double-click on each of the three major rivers in Siberia. If the map that appears after you select a river has that river highlighted in yellow, you are on the right track. With each of the maps there will be a brief description of the river and its environs that you may access by selecting *Geography* at the left-hand side of the screen. As you review these narratives you should be able to answer the following questions. As you read about the geography of the rivers, be sure to click on *Related Topics* and select *Longest Rivers, Asia*, to see how these rivers compare to other watercourses.

9. How does each of these three rivers rank in length as compared to other rivers in Asia?

10. Which of these rivers is usually frozen and non-navigable from November through May?

11. Which of these rivers begins near Lake Baikal and empties into the Laptev Sea?

12. Which river has as its chief tributary the Irtysh and has its source in the Altai Mountains? What types of products are usually transported on this river?

Keywords: **Russia, Siberia, rivers, folk music**

Chapter 3: North America
Uh Oh, Canada: A Fractured Federal Fairy Tale

Canada faces many challenges from within and beyond its borders. Among the most difficult issues facing Canadians today comes in the form of internal problems fueled by ethnic separatist movements by two groups---Inuit peoples of Nunavut and the Québecois of Québec. In addition to challenges to hold the provinces together, Canada must also contend with the overwhelming cultural and economic influences of its neighbor the Unites States. United States residents outnumber Canadians nearly ten to one. Since the vast majority of Canadians live within 100 miles of the U.S. border, it is nearly impossible for Canadians to escape U.S. media and cultural influences. If Canada survives its struggles for identity and unity in the late 20th century, what awaits this decidedly multicultural federation in the next millennium? For a closer look at the two groups that pose the greatest threat to Canada's federal integrity, let's take a virtual tour of Québec and Nunavut and see how the goals of each figure into the future of the country.

PART I: Québec
In 1995, Québeckers narrowly defeated a referendum to separate from Canada. What cultural and political forces could bring about such a drastic measure? In comparison, if you live in the United States, how do you think you would feel if Iowa or California decided to secede from the union? What happened the last time states tried that option? While civil war does not appear to be imminent in Canada, the consequences of Québec separating from the country would be dramatic. To learn more about Québec and why a large number of its residents believe a separation from Canada is in the best interest of the province, first create a map of the country using the *Find* tool. Once you have a map of the province on the screen, click on *Geography* to read a short description of Québec and answer the following questions.

1. From a purely economic perspective, what would Canada lose if Québec were to separate from the rest of the country?

2. What event in 1974 was an attempt to preserve French culture in Québec? What has happened since 1974 to change the political climate in the province? Do you think Québec will break away from Canada---why or why not?

3. Given that 80% of the Québec population claim some sort of French ancestry, why do you think only 60% speak French as their first language? Why would Québeckers consider language as such a critical issue in their provincial identity? Can you think of any parallel situations in the United States----states where language has become a volatile political as well as cultural issue?

PART II: And They Said We'll Have Nunavut
The struggles of the Québecois to retain and protect their cultural identity are well known. But what of another influential minority group within Canada---the Inuit? The Inuit are among 350,000 of Canada's First Nations (who, in the United States, are called Native Americans) a group that also include Mohawk, Cree, Algonquin, Huron, and others. Of these peoples, the Inuit have been politically successful in securing a homeland while most others live on

8

reservations throughout Canada. In 1999, the Inuit took control of a large section of the Northwest Territories known as Nunavut. To create a map of this new part of Canada, use the *Find* box and type in Nunavut. Check out the brief description of the territory by clicking on *Geography* and answer these questions.

4. What and where is the capital of Nunavut? Can you think of any other province or state whose capital is not on the mainland?

5. What types of natural resources are found in Nunavut? What might be some of the problems encountered as the Inuit attempt to exploit their mineral wealth?

6. What do the Inuit and the Québecois have in common---or do they share any common issues?

* Web Link Activity: Exploring Nunavut
Click on the *Web Links* for Nunavut and select "Exploring Nunavut" and go to the "The People" link. At the bottom of that page you should see a list of topics---select "Inuit culture" and read the narrative about "The Baffin Region" by Ann Meekijuk Hanson and "The Kivalliq Region" by Peter Ernerk. Read through this informative section on Inuit culture and answer the following questions.

7. Translate or describe the significance of the meaning of the following Inuit words:

a. Isomainaquiijutitt:

b. Aalu, misiraq, and Nirukkaq:

c. Inuit:

8. How do the Inuit go about naming landscape features---give an example.

9. Not only does Peter Ernerk dispel the myth of the Inuit "kiss" but also of the notion that there are many words for snow---in fact there is only one, *aput*. He also suggests that you would do well not to call an Inuit an "Eskimo"---why?

Keywords: **Canada, Québec, Québecois, Nunavut, Inuit, culture, language**

*This Web Link Activity is based on a web site available to the authors via Encarta® and Microsoft® at the time of publication. The authors and publisher do not control the actual availability of the web site, and apologize for any inconvenience if the web site is not available to you.

Chapter 4: Middle America

The Land of Shake and Bake: Landscapes of Earthquakes and Volcanoes in Mexico

Would you locate a city on an almost unbuildable location because of your belief in a myth? The Aztecs did. The site for their capital of Tenochtitlán was selected because legend had it that wherever an eagle with a serpent in its beak landed on a cactus would be the ideal mystical place to site their capital. Unfortunately, that cactus was growing on not the most attractive real estate--on a reed and mud island in the middle of a large fresh water lake (Lake Texcoco). The symbolism of the eagle with the serpent on a cactus is still a powerful one in Mexico; it is depicted on their peso coin and on their flag.

Present day Mexico City is located right on top of the Aztec capital. In fact, when digging for an extension of Mexico City's subway system, workers discovered the base of an important ceremonial pyramid from the Aztec era. That stonework has been preserved next to a modern Mexican marketplace and a Spanish colonial-era church at the Plaza de los Tres Culturas (Plaza of the Three Cultures—Aztec or Indian, Spanish and modern Mexican). Each of those three cultures has shaped and reshaped this modern, bustling city in central Mexico. Locate this important Plaza by using the *Find Places* command and bringing up a map of Mexico City (Museo de la Cuidad de México scale).

1. From this map select two place names that are in the language of the Aztecs and not the language of the Spanish conquerors.

While the site of Mexico City may have been ill-advised, the situation (i.e., its location relative to important trade routes; its centrality as a hub of communication and transportation within the nation) is unparalleled.

How ill-advised was the site of present day Mexico City? Although the important tourist attraction of the floating gardens of Xochimilco is about all that remains of ancient Lake Texcoco, the legacy of locating a city on a lake bed of rather recent sedimentary deposition remains. While still focusing on Mexico City, choose *Sights and Sounds*. There you will see the "Sinking Palace" (i.e., The Palace of the Fine Arts building) as it has sunk over 15 feet since construction due to draining of the underlying aquifer for municipal water needs. It is said that the foundation of the Latin American Towers Building (over 41 stories in height and, at one time, the tallest building in Mexico) extends almost as deeply below ground in order to reach the firmer bedrock beneath the lakebed sediments.

Unfortunately, not all buildings in rapidly growing Mexico City (the most populated and polluted city in North America) are as soundly constructed as the Latin American Towers. Why is this a problem? Find Mexico by typing it in the box and choose the *Map Style* labeled *Physical* and the extension labeled *Tectonic*. Tectonic activity is mountain building activity commonly associated with earthquakes and volcanoes. Mexico City has both nearby. Earthquakes have had a devastating impact on Mexico City. In 1985, the strongest quake recorded in recent years (an 8.2 on the Richter scale), had an epicenter some 200 miles from

Mexico City. Because of the inherent geological instability of the lake sediments upon which Mexico City stands, the shock waves from that earthquake leveled many buildings in Mexico City, killing and injuring thousands of people. It was as if Mexico City were sitting in a bowl (i.e., the Basin of Mexico) filled with Jell-O (i.e., unconsolidated lake sediments).

2. According to the *Tectonic* map, how many crustal plates are near Mexico City? Name them. Is it any wonder that the city has received and continues to receive many tremors?

3. What part of Mexico is even more earthquake-prone than central Mexico?

4. Click on *Sights and Sounds* of Mexico City (Districto Federal scale) and focus on the view of Mexico City Skyline presented there. What do you notice in the distance? That's right. There are two snow-covered mountains beyond the lower mountains that form the Basin of Mexico. These two mountains are of volcanic origin.

Click once again on the *Find* function and look up the information for Volcan Iztaccihuatl and Volcan Popocatepetl, which together have been declared a National Park. These place names are long and difficult for us to pronounce because they are in the language of the Aztecs, not that of the Spanish conquerors.

5. If one were staging a campaign to save the *teporingo*, what would one be saving?

6. How does *Iztaccihuatl* translate? (Hint: Look under *Country Contents* for Mexico and click on *Geographic Features*. Now slide the slide bar down until the name of the volcano appears). Do you think that it is an appropriate place name? Why or why not?

Keywords: **Mexico, Mexico City, earthquakes, volcanoes**

de Blij and Muller, *Geography: Realms, Regions, and Concepts*, 9e
Chapter 5: South America
Suriname: Going Dutch in the Rain Forest

The dominant colonial forces for the majority of South America were the Spanish and the Portuguese, except on the northeast coast of the continent in a region known today as the "Three Guianas." The Three Guianas include Guyana, French Guiana, and Suriname. What makes the Three Guianas unique in South America is that they are more similar in character to islands in the Caribbean than with their "Latin" American neighbors. Of the three, only French Guiana remains under colonial rule as a French *departement*. Guyana was ruled by the British and gained independence in 1966. Suriname was a Dutch colony and has been an independent state since 1975. Of the three, we will focus on Suriname.

Suriname has a diverse cultural heritage as a result of European, African and Asian immigration with many of those belonging to the latter two groups--descendants of slaves or indentured servants. In addition to social complexities, Suriname faces serious economic and environmental challenges that are highly interrelated and difficult to reconcile given the country's political and monetary instability. For an introduction to Suriname, create a map of the country using the *Find* box. When the map appears, click on the country's name and select *Sights and Sounds* and then *Society*. View the slides, read the narrative and answer the following questions.

1. What are some of the examples of remaining vestiges of Dutch culture in Suriname?

2. What are some of the long-lasting economic impacts of Dutch colonization?

3. What is the official language of Suriname?

4. What is the most widely spoken language in Suriname and how does this reflect cultural integration and the history of the country?

Another way to learn about this complex country is to explore the *Web Links* for Suriname. First, let's visit the Lonely Planet's *Destination Suriname* by clicking on the country name and then selecting *Web Links*. Read through each of the sections and consider the following questions.

5. One of things that may have surprised you about Suriname is the strong presence of Hinduism. How did Hinduism and South Asian culture diffuse to Suriname and how does it influence the country today?

6. After reading about travel in Suriname, do you have any interest in making this country a vacation destination—why or why not?

Next, check out the *Tropical Rainforest in Suriname* link. Select the *Rainforest* site and prepare to take a multimedia tour of the Suriname rainforest. There are a variety of images and sounds on this site so take advantage of them as you take a virtual tour of Suriname and its rainforest. After reviewing the web page, you should have no trouble answering these questions about Suriname in the surrounding region.

7. The lack of transportation infrastructure affected economic development in the Three Guianas. Could you drive your car from Paramaribo to Caracas? Why or why not?

8. What is the *lingua franca* of Suriname, what are its roots and when did it develop?

9. What does *a swampen en ritsen* in Suriname have in common with the landscape of the Netherlands?

10. How is the savanna different from the coastal plain?

11. What sort of houses do Amerindians build?

12. What type of "brew" might you enjoy with Galibi smoked fish?

13. Bauxite has long provided Suriname with a major export product, but how has aluminum refining affected Maroon villages in the rainforest?

14. Who are the Maroons and what role have they played in our understanding of rainforest ecosystems?

15. One of the major reasons why the rainforest is threatened has been the exhaustion of once extensive bauxite deposits. The Suriname government has shifted its economic development emphasis from bauxite mining to logging and gold mining. What have been the environmental consequences of these alternative economic initiatives?

16. What types of flora and fauna do you find in the Suriname rainforest?

17. What type of plant could be described as the rainforest's "trash can?"

There is much more to learn about Suriname. Be sure to check out other *Web Links* as well as *Facts and Figures* to increase your understanding of this unique country and region within South America. And watch where you step. There might be a *Teraphusa leblondi* in your path!

Keywords: **Suriname, Three Guianas, rainforest, colonialism, Amerindians**

Chapter 6: North Africa/Southwest Asia
Are Kurds in the Way?
The Political Geography of Kurdistan

Before you begin this activity, study closely the map displayed in Figure 6-12 (p. 301 in the 8th edition; Figure 6-13, p. 309 in the 9th edition) and the accompanying narrative on a Kurd homeland (p. 302 in the 8th edition; p. 309, in the 9th edition). It is clear from the book's discussion that there are major differences between a nation and a state.

1. In your own words, state what these differences are.

We have learned that the Kurds are "fractious and fragmented" (p. 302), but who are they really? What is their ethnic, cultural, religious and linguistic background? Why, like the Basques on the border region between Spain and France, do they want their own autonomous state?

Although the textbook introduces us to the Kurdish people and their plight, we really don't know much about the Kurds themselves. Using the *Find* command, bring up any material that might exist on the Kurds.

2. Where is their traditional cultural hearth located? Hint: Note the slide of Lake Van (Van Gölü) in eastern Turkey in *Sights and Sounds*. The lake is saline and contains many other minerals as well. This area appears to be the traditional core area of their culture realm. Look at the map in the textbook of the disputed region (p. 301 in the 8th edition; Figure 6-13, p. 309 in the 9th edition).

The Kurds traditionally lived in rather isolated mountainous villages in the area now occupied by portions of Iran, Iraq, Turkey, Syria, and Azerbaijan.

3. Why is the 36th parallel shown on the map in the Kurdish area of Iraq?

Before the Gulf War, Kurds were located throughout the Iraqi region of Kordestan. Click on Kordestan and find out what makes the region so valuable.

4. In the description of Kurdistan, what is mentioned specifically about the Zagros Mountains which form the southeastern border of the region that the Kurds would like to make their autonomous homeland?

5. How important is this resource in making the states that have significant Kurdish minorities reticent to give up land to create such a new state?

14

6. Do the Kurds have their own separate language and culture? Click on *Sights and Sounds* and take an "Afternoon Break" in Sanli Urfa, Turkey.

7. Now visit the Kurdish Village in *Sights and Sounds*. Both the Iranians and the Kurds are adherents of Mohammed the Prophet. How, then, do they differ in religious practice? See if you can find a site on the Microsoft® Encarta® CD-ROM or a suggested Web site that would briefly explain the difference between the Sunni (Kurdish) and Shi'ite (Iranian) branches of Islam.

8. What happened in 1979 to make the Iranians more hostile to the demands for Kurdish autonomy? (Hint: check for *Society* under the heading Iran)

9. Of the 16 million estimated Kurdish people, how many live in Turkey?

10. What other language in the region is closest to the Kurdish language?

Now click on the website for Kurdistan. Then click on the title "Who Are the Kurds?" Answer the following questions.

11. What are Kurds in Turkey known as?

12. What is the Treaty of Algiers (1975) and why is it fateful to the Kurds?

13. According to Kurdish accounts, what happened to the city of Halabja in March 1988?

Keywords: **Kurds, Iran, Iraq, Turkey, state, nation**

Chapter 7: Subsaharan Africa

The Slow Plague:
The Spread of AIDS in Central Africa

There is nothing lighthearted or whimsical in this title. It is, in fact, the title of a book on the subject by geographer Peter Gould. The World Health Organization (WHO) estimates that of the 18 million reported cases of AIDS (acquired immune deficiency syndrome) in the world, 11 million have been reported in Africa alone. Even more frightening are estimates of the number of African men who test HIV positive. In some areas that percentage is almost half of all adult males. Central and equatorial Africa has always been an area where potentially deadly diseases are found in abundance—river blindness, schistosomiasis, malaria, and trachoma to name a few.

Imagine how much more deadly these diseases are among a population with already weakened immune systems due to poor diet and lack of access to adequate medical facilities. In order to find out more about the spatial distribution of AIDS in Subsaharan Africa, click on *Statistics Center*. Under the category of *Health* statistics, scroll down until you find *acquired immune deficiency syndrome (AIDS) statistics*. At the bottom of the screen create a *Statistical Table* first. It would be best to sort the table by Rank where it asks for *Sorting Method*.

1. Of the thirty countries with the greatest number of reported cases of AIDS, how many of them are in Subsaharan Africa?

2.a Of the three countries in Subsaharan Africa that are in the top five countries in the world as measured by the number of reported cases of AIDS, what percentage of the total population do these numbers represent? (Divide the number of reported cases of AIDS by the total population of the countries involved. *Population, total* is a category under the more general *Population* category. You may round your figures to the nearest one-tenth of one percent).

2.b How do these figures compare with the proportion of the general population that been infected with the AIDS virus in the United States?

Perhaps the main difference in the spread of AIDS in Africa vis-à-vis the United States (and much of the rest of the developed world) is that the disease in Africa is mainly spread within the heterosexual population. In the United States, the disease is spread mainly in two ways: 1) by unprotected sex among high risk groups including homosexuals and heterosexuals with multiple sex partners; or 2) by infected blood obtained from tainted blood transfusions, passing the disease from infected mother to child or by the sharing of infected needles among intravenous drug users.

Using the *Find* command, type in 'Africa' and construct a map of the number of reported AIDS cases in Africa.

3. Does there appear to be a core area from which the disease might have emanated and spread?

4. Lake Victoria has always been a shared resource and a means of water-borne transportation and communication among the former British High Commission territories in East Africa (Kenya, Uganda and Tanganyika—now Tanzania). Do you think it has played any role in the spread of AIDS in East Africa?

The World Health Organization has tried to target the populations in Africa for educational programs and preventative measures to stop the spread of AIDS that it feels are the most at risk. Among these are hotel workers, prostitutes and truck drivers on the trans-Africa trucking routes passing through Kenya and countries in central Africa.

And yet there are countries in Africa with very few reported cases of the AIDS virus.

5. Where are the countries in Africa located that appear to be at the lowest risk of acquiring the HIV virus that causes AIDS?

6. What factors can you propose that would account for these place-to-place differences in the spread of this deadly disease?

7. The description of AIDS within the *Details* section at the bottom of the *Statistics Center* suggests that perhaps only 15 percent of the cases of the AIDS virus are reported in the official statistics. Why would a country wish to withhold that kind of information from statistical reporting sources such as the Centers for Disease Control (CDC) or the World Health Organization (WHO)?

Keywords: **AIDS, HIV, diffusion, Africa**

Chapter 8: South Asia

The Untouchables: A Caste of Millions

Hinduism is an ethnic religion practiced mainly in India, though it has diffused throughout the world mainly through immigration. Hindus do not seek converts as do other universalizing religions such as Islam and Christianity. Hinduism is closely linked to social stratification within India in large part because of the caste system. Castes are a Hindu system of social rank based on ancestry, family relationships, and vocation. Hindus believe in reincarnation and thus the caste into which you are born is reflective of your actions in a previous life. The most prestigious caste members are the Brahmans and the lowest caste is the Untouchables. Traditionally, Untouchables have been the poorest of the poor, ostracized socially, politically, and economically by upper caste members. The plight of the Untouchables compelled British colonial rulers and subsequent Indian governments to attempt to improve the lives of millions of the lowest caste members. Recent gains by the Untouchables can be attributed to political activism spurred by increased voter registration and social reform.

PART I: Hinduism and Indian Society

In this activity we will take a virtual trip to India and learn more about the untouchables and Indian culture. The first thing we need to do is create a map of India using the *Find* box. Once you have India on the screen, click on *Society* and read each section keeping in mind how religion may play a role in different aspects of life. Be sure to click on the slides that accompany the narratives and then answer the following.

1. How has the interpretation of the caste system changed over time?

2. Given India's political and economic climate, what do you see as the future for the Untouchables?

3. Although this might be difficult to do given the great cultural differences between India and the United States, can you think of any groups of people in the U.S. that some might consider to be Untouchables? What are some of the parallels between India's Untouchables and some of America's poor and disadvantaged?

4. In recent times, there have been a number of violent conflicts between Muslims and Hindus. Why are these two religious groups at odds with each other?

5. How does gender figure into Indian culture? Do women have their own "caste" within society as a whole and if so, give some examples.

PART II: Diffusion of Hinduism

India is not the only Asian country with a large percentage of its population as practicing Hindus. Hinduism has spread from the subcontinent to bordering countries and beyond. To identify this diffusion pattern, go to each country that borders India including Sri Lanka and click on *Facts and Figures* to determine what percentage of that country's population are Hindus. After you have this information, respond to the following.

6. List India's neighbors and the percent of Hindus for each.

7. Which country actually has a higher percent of its population as practicing Hindus than India? Go to a map of this country and click on *Sights and Sounds*. Pay special attention to the slide titled "Caste System of the Newar" and also listen to the "Gaine Music" audio sample. Who are the Gaine people and how does this caste differ from the Newar?

Hinduism has diffused far beyond the borders of India. For some examples of how far Hinduism has spread throughout the world, go to each of the following places and then click on *Sights and Sounds* for each (Hint: you can use this shortcut to find these pictures---enter the country name in the *Find* box then click on the slide title in *Content* and go directly to item you are searching for).

Mauritius…view the slide titled, "Hindu Dominance"
Bali…click on "Seaside Offering"
Fiji….look for "Indian Shopkeeper in Fiji"

After viewing these images and reading the captions answer the following questions.

8. How did Hinduism diffuse to each of these locations. Has the relocation of Hindus to any of these places caused any cultural conflicts? If so, please elaborate.

9. Do you think Hinduism plays the same role in these countries as it does in India? Why or why not?

Keywords: **India, Hinduism, caste system, untouchables, diffusion, religion**

Chapter 9: East Asia
Dragon China into the 21st Century

There seems little doubt that the greatest economic power of the 21st century will be the People's Republic of China. It could hardly be otherwise. The United States has only 1/4th the number of people and its rate of economic growth, the envy of many developed nations in Europe and elsewhere, is much slower than that of China. On the other hand, China has a long way to go. It has come from a peasant society under the thumb of feudal warlords at the beginning of this century through the worst "cleansing" excesses of the Communist regime--the Red Guard, the "Gang of Four" and the cult of Mao--to what it is today. China is still ostensibly a Communist regime, but is poised to become a major player on the world's <u>economic</u> stage if not its political equivalent.

China by no means comes close to the egalitarian ideal of a socialist worker's paradise. There are vast differences in the income levels among the provinces. Those to the south, especially those near the newly acquired "province" of Hong Kong, are growing very rapidly as are the coastal provinces along the Pacific Rim. Many interior provinces, on the other hand, suffer great poverty and practice much more of a traditional agricultural peasant economy.

One of the brightest stars on China's horizon and the nexus for a huge infusion of capital both foreign and domestic is the city of Shanghai and its surrounding region. Shanghai has always been the nation's largest and most important city when it comes to foreign trade. With a navigable deep-water port facility and excellent supporting infrastructure, Shanghai is truly China's gateway and entrepôt to the rest of the world. Click on *Find* and bring up the map of the area around Shanghai.

1. What factors contribute to the excellence of its site and situation?

If Shanghai is considered together with the growing city of Pudong on the opposite bank of the Huangpu Jiang River, it has been said that almost 25 percent of the world's construction cranes are located there. What are they building? To find out more about Pudong, click on 'Shanghai' or 'Pudong' and then click on *Geography*.

2. What does the Chinese government have in store for Pudong in the future? Why is this city, referred to in the 1920s as the "Paris of the East", re-experiencing such phenomenal growth?

In the 1920s, the Bund, a governmental and shopping district near the waterfront, was the place to see and to be seen. Click on *Sights and Sounds* to see for yourself "Shanghai Shoppers on the Bund". The legacy of the "Golden Age" of European power in this area of China can be seen it the ornate façades of the buildings along the Bund and elsewhere. For a photographic display of some of the best examples of turn-of-the century European colonial architecture which appears somewhat incongruous when compared with indigenous traditions of building in China, click on the *Web Links* and examine the site called "Shanghai's Historical Western Architecture". Don't worry if it takes awhile to download the files. The stunning black and white photographs, taken all around the city of Shanghai, are worth the wait.

Find out a little more about this fascinating city. Using the *Find* box, type in 'Shanghai' and when a map of the environs is on the screen, click on *Geography* where there is some interesting textual material about Shanghai.

3. What facilitates the movement of agricultural produce from the farms in the rural part of the municipality of Shanghai to the city center?

4. What controversial treaty opened up Shanghai for foreign trade?

5. In addition to the rural Chinese seeking employment and a better life in Shanghai, what other refugees that we do not normally associate with China did Shanghai attract?

6. The Chinese actress Jiang Qing got her start in the film studios of Shanghai when the city was known at the "Paris of the East." By what other name do we know this person today?

7. What park in central Shanghai was laid out during the Ming Dynasty (1368-1644 AD)

8. What is the Chinese name for the Bund?

Now click on *Sights and Sounds* and focus your attention on the last slide in the sequence.

9. What has the Chinese government done to preserve ancient crafts such as the carving of jade and other semi-precious stones?

21

Chinese financiers living overseas (i.e., Chinese expatriates), make much of the foreign investment in Shanghai. These are Chinese people whose families fled mainland China when the Communists took over in 1949. This huge migration, by some estimates the greatest the world has ever known, is sometimes referred to as the Chinese diaspora or scattering. They reestablished themselves throughout East and Southeast Asia and around the world. The hard driving economies of Singapore and Hong Kong, experiencing the first real downturn in their economies at the time of this writing, are largely controlled by these overseas Chinese expatriates. Persons of Chinese extraction from as far away as Vancouver or New York are flocking to Shanghai to invest in real estate ventures, start new businesses and build the industrial and service infrastructure necessary for Shanghai to emerge as a truly global city to rival Tokyo, London or New York. The investment is often made through tight-knit family organizations called *guang-xi*. These investment groups are largely closed to outside investors which has created suspicion and jealousy among entrepreneurs who feel that they are being systematically excluded from some lucrative opportunities. They would argue that to be of Chinese ancestry gives one a leg up in this globally competitive Chinese market.

10. Do you think that such family-owned business arrangements are much different than limited partnerships in the United States and elsewhere? Why or why not?

Keywords: **China, Shanghai, guang-xi, Bund, Pudong, overseas Chinese**

A Close Shave:
Will Myanmar Ever Revert to Burma?

Myanmar (Burma) is one of the poorest, least Westernized countries and is run by one of the most oppressive military regimes in all of Southeast Asia. On the other hand, it is one of the most spiritual and beautiful places on Earth. Myanmar has consciously turned its back on the modern world and cut off all contact with the West. In turn, the military regime that came to power in 1988 (the State Law and Order Restoration Council—SLORC) continues to defy the will of the Burmese people. At first, it was thought that this military regime might right the corruption that was rampant throughout the country. Bribery and scandal was even tainting the Buddhist monks. Elections were called for in 1990 and, despite there being 93 parties running, many representing small ethnic minority groups that live near the border regions with China and Thailand, the National League for Democracy (NLD) won an overwhelming majority of the vote. But, the military regime refused to recognize that the people had spoken and jailed many of the members of the NLD.

Click on 'Myanmar' or 'Burma' using the *Find* tool. Go to *Web Links* for Burma and examine the **"Destination Myanmar"** travel guide prepared by the British Lonely Planet travel agency. They call Myanmar "a bizarre, inept Orwellian society that has withdrawn from contact with the late 20th century". Although George Orwell actually did write a book based on his experience in Burma, it is likely that the quotation refers to his more famous novel *1984*.

1. In your opinion, do you believe Myanmar fits this description as an Orwellian society? Why or why not?

Now, go to the web site entitled "Project Burma," which keeps track of articles referring to human rights abuses in Myanmar, and focus on the section dealing with Burma-US Relations.

2. Who is the outspoken advocate for civil rights in Burma that won the Nobel Peace Prize for her efforts?

It may seem incongruous that such a politically repressive society is also a very spiritual one. The focus of most lives is the Buddhist religion and the center of a settlement's life is the monastery and the pagoda(s) dedicated to The Enlightened One. The main pagoda in Yangon (Rangoon), the capital and largest city, is covered with gold leaf. That Shwedagon Paya pagoda (temple) is alleged to contain eight hairs from the head of the Buddha. The form of Buddhism practiced in Burma differs from that practiced in Tibet or in China.

Using *Web Links* examine the subject of religion in "Destination Myanmar".

3. What type of Buddhism is practiced in Burma?

Next to petroleum and natural gas, Burma depends on tourism as its most important source of hard currency. The year 1996 was even declared the year to Visit Mynamar in official governmental policy. It is sometimes interesting to compare the tourist attractions felt worthy of a visit in an official source with those deemed viewable in a more "objective" non-governmental source. That's what we'd like you to do. Click on the *Web Links* for "Welcome to Myanmar", an official source of governmental propaganda on tourism. Then examine "Destination Myanmar" developed by the travel group called Lonely Planet. Both are colorful and filled with pictures. One of the funniest is found in Lonely Planet's web site. The photograph is called "Cheroot vendor".

Just what do you suppose that woman is smoking!?

At that web site, you will find out that the Burmese like to drink tepid weak green tea and chew betel. What exactly is betel?

Practically all of the tourist sites listed in the official site are religious pagodas, many associated with ancient capitals or centers of earlier civilization. None of the buildings left from the British colonial era, not even the famous Strand Hotel in Rangoon (Yangon), are mentioned in the official tourist source. Seemingly more important are ancient capitals.

4. What is the name of the ancient capital on the eastern bank of the Irrawaddy (Ayeyarwaddy) River, a city of four million pagodas?

5. This city was declared the capital by King Mindon in 1857. What is the name of the city and what famous British author made it famous?

The official guide implies that these tourist destinations are easy to get to by air or road. The Lonely Planet is more sanguine. Burma doesn't have an all-weather road that traverses the country from north to south. Internal air travel is dicey with the notion of keeping to a schedule somewhat of a foreign notion. If visitors want to go upcountry away from the major tourist sites, they must have the consent of the government. No cruise ships are allowed into Myanmar at all. Travel in Burma appears to be for the adventurous.

As befits a lesser-developed economy, over seventy percent of the gainfully employed labor force are in the extractive industries—agriculture, forestry, fishing and mining. While rice is the main agricultural staple, one should not overlook the fact that 60 to 80 percent of the heroin in the United States originates in the Golden Triangle portion of Myanmar at the border with Thailand, China and Laos. The farmer receives only about $700 a kilo for the unrefined heroin derived from opium poppies. By the time it is refined in Bangkok, Thailand that kilo will be worth $10,000. And when that kilo hits the streets of New York City its value will balloon to $750,000.

Scroll down the Lonely Planet's information about Myanmar until you come to the Heading marked 'Warning!'

6. What's the warning about?

7. Who controls the Golden Triangle portion of Myanmar northeast of Mandalay near the Thai border?

8. How many troops does this person control?

Keywords: **Myanmar (Burma), Yangon (Rangoon), Mandalay, Buddhism**

Chapter 11: The Austral Realm
Australia Rocks!

Diamonds from the Kimberly Plateau aren't the only rocks that you will find in Australia. While rich in mineral wealth, Australia could draw on a variety of "rocks" to symbolize its storied history and diverse culture. Now mates let's take a "rock tour" of this country/continent and find out what makes Australia really rock. The first thing you need to do is create a map of Australia using the *Find* box. Then, for each of the following, go to the location needed to answer the associated questions.

PART I: Rock That Doesn't Roll: Uluru Rock
Uluru Rock, formerly known as Ayers Rock, is a major tourist attraction in the Northern Territory and is considered sacred space by the Aborigines. The massive sandstone monolith is breathtakingly beautiful especially as its color varies throughout the day and appears scarlet by sunset. To become a virtual tourist and visit this site, use the *Find* option and enter Uluru and select Uluru National Park in the *Places* box. Once you have a map of Uluru National Park, take a few minutes to read through the *Geography* narrative and view the Rock via *Sights and Sounds*.

1. What do Uluru and Kata Tjuta have in common?

2. Why do you think the Australian government returned the land that is now the Uluru National Park back to the Aborigines?

3. Do you think the United States government would ever offer a similar deal to Native Americans? Explain why you feel this way.

PART II: Australian Rock that Rocks
From a red rock to blues-rock. In Alice Springs, 325 kilometers (200 miles) northeast of Uluru National Park, the Central Australian Aboriginal Media Association has established a recording studio to preserve and foster appreciation for traditional and contemporary aboriginal music. Return to *Sights and Sounds* for Uluru National Park and listen to the audio sample of aboriginal rock music.

4. What events have brought about a movement to preserve and promote aboriginal music?

5. What did you think of the music sample by the Titjkala Desert Oaks Band? Do you think you or your friends would buy their CD? Why or why not?

Australia is also home to other rock musicians. It may be hard to imagine but rock music is an Australian export commodity just like wheat, wool, or beer. Exports of Australian music exceeded $200 million (Australian) annually during the 1990s. Among the country's more internationally known rockers---AC/DC, Air Supply, Crowded House, INXS, Little River Band, Men at Work, Savage Garden, Silverchair, and Rick Springfield. The Little River Band is the only group in this list to have a geographical name. The Little River (also known as the Boyd River) is located in New South Wales. Of course, we could include Rick Springfield in our geographical rocker category. There are at least four towns named Springfield in Australia (but none named Rick). Listed below are four "Springfields." Select one, follow the activities and answer the corresponding questions.

Select one and circle your choice:
Springfield, New South Wales, Australia
Springfield, Queensland, Australia
Springfield, Tasmania, Australia
Springfield, Victoria, Australia

6. Locate the Springfield of your choice on a map of Australia by using the *Find* box. When you have a map of "your" Springfield, you will be ready to rock and roll with these questions. You will be using *Map Styles* and *Legend* to assist with your answers.

a. What is the approximate *Population* of your Springfield?

b. In which *Ecoregion* is your Springfield located?

c. What is the *Climate* regime for your Springfield?

d. What is the *Average Temperature* in July in your Springfield?

e. What is the *Annual Precipitation* for your Springfield area?

f. Does your Springfield show up on the map of *Earth by Night*? If not, which Australian cities do?

g. What do you suppose people in this Springfield do for a living?

Keywords: **Australia, Uluru Rock, Aborigines, Rock Music**

Chapter 12: The Pacific Realm

Want to Know Samoa? Read On!

If you have ever taken a geography place location test and it required you to identify some of the islands sprinkled in the western Pacific Ocean, you probably were learning of many of these places for the first time. After all, how often do many of us get to travel to such far-flung places as Samoa and Fiji? While these islands are often thought of as exotic vacation destinations, there are many intriguing facets to their cultures and economies that deserve our attention. In this activity we will go island hopping to Samoa and Fiji to sample their geographies.

PART I: Samoa

Samoa is among the least industrialized economies in the world and struggles each year to cope with a serious trade-deficit. The economy is largely dependent on the primary sector activities including timbering and agriculture. What Samoa lacks in advanced economic development it makes up for with a rich cultural heritage that can be marketed to tourists.   To take a virtual tour of Samoa, use the *Find* option and create a map of this tiny Pacific island country. Click on *Society*, read each section and answer the following. Be sure to take advantage of the additional information such as *Related Topics*, slides, and the *Video Clip* at the beginning of the discussion.

1. What is the dominant ethnic group in Samoa and what is the language spoken by most people?

2. Describe the typical Samoan village.   How is this different from a typical American small town?

3. What is the difference between a *matai* and the *Fono*?

4. If you were a dinner guest for an evening at a Samoan house, what might you expect to eat and how should you behave to feel welcome?

5. What is the focus of the "Swarm of the Palolo?"

6. How would you describe the transportation and communications infrastructure in Samoa?

Let's take a closer look at Samoa by visiting Apai, the capital. To go to Apai you may use either the *Find* option or simply use your computer's mouse and click on the capital on the map of Samoa you already have on your screen. When you have a map featuring Apai, select *Geography* to find the information to answer the following.

7. Which two world powers fought for control of this region during the late 1800s and how did the weather play an important role in ending the conflict?

8. What type of manufacturing would you find if you visited the Apai area? Why do you think such products would be fabricated in Samoa?

28

PART II: Fiji

Can you name an international sports celebrity from Fiji? Have you ever heard of Vijay Singh, a major star of the professional golf tour? He may very well be the most famous person from Fiji in contemporary culture but even it's a long way from Fiji to Pebble Beach. Imagine you are spectator at a golf tournament and have an opportunity to meet Mr. Singh. What would you say? Perhaps you might want to impress him with your knowledge of his native Pacific region? To prepare yourself for this encounter, let's take a short course in Fijian culture. First, create a map of Fiji using the *Find* box. Next, review the section on *Society* and take a brief visual tour of the islands by clicking on *Sights and Sounds*. Now you are ready to answer some basic questions about Fiji.

9. What is the source of ethnic tensions in Fiji and what has transpired since the 1980s to increase tensions between these two groups?

10. How would you describe the gender roles in a typical Fijian household?

11. Mr. Singh may be disappointed but golf is not the national sport of Fiji---what is? What game is often played by women and what are the rewards to the winning team?

12. What are some similarities and differences you have observed between Fijian and Samoan culture and why do you think these characteristics exist?

If you were able to answer those questions you are ready to tackle an even more challenging task---learning the Fijian language. Go to *Web Links* and select "Common Fijian Words and Phrases." Now, using this language page as your guide, translate this hypothetical conversation between you and Mr. Singh as you follow him to the 18th tee.

Mr. Singh:	Good morning.
You:	**Ni sa yadra turaga**.
Mr. Singh:	You look like a brilliant geographer to me. So tell me, do you know anything about Fiji?
You:	**Vakalailai.**
Mr. Singh:	Could you locate my hometown of Lautoka on a map?
You:	**Io!**
Mr. Singh:	Pardon me, but you are sitting on my golf bag!
You:	**Tulou!**
Mr. Singh:	That's okay. Tell me, how many large islands comprise Fiji?
You:	**Rua.**
Mr. Singh:	How many rooms are there in a traditional Fijian "burus?"
You:	**Dua.**
Mr. Singh:	Why, you *are* a brilliant geographer!
You:	**Vinaka vakelevu!**

Keywords: **Samoa, Fiji, Pacific Ocean, islands, ethnicity, language**

de Blij, *Human Geography: Culture, Society and Space*
Part One: Environment and Humanity
Rocky Mountain High (and Corn Belt Low)

Thirty years ago, Peter Gould, a geographer at The Pennsylvania State University, introduced the notion of **mental maps** into the geographic literature. Using a sophisticated mathematical procedure known as factor analysis, he was able to produce isoline maps of residential preferences of students attending different universities in the country including his own. Figure 1-9 in your textbook provides two examples—the upper map is the view from California (specifically UCLA in Los Angeles) and the lower map is that of students at The Pennsylvania State University. The question that Gould asked of these students allowed them to really "blue sky" their responses as if they didn't have a care in the world other than choosing the most desirable residential location possible. The method used by Gould to extract the communality in the responses of the students has been criticized, but the maps that are produced continue to fascinate.

Most students had a strong preference for their home state irrespective of the fact they were told they could disregard family obligations. Gould called this the **local dome effect.**

Secondly, there was a general **national trend** in evidence. In the late 1960s, students preferred California—it was usually selected second only to the home state. The residential desirability surface then followed the topography of the country in a manner of speaking. The Great Basin states (e.g., Nevada, Utah) were as low ranked as their relative elevations. The desirability surface then rose as the Rocky Mountains loomed to the east but fell again beyond the front range of the Rockies and remained low all across the agricultural Great Plains and Corn Belt states. The residential desirability surface then rose toward the Middle Atlantic States and those along the Eastern Seaboard until the Mason-Dixon Line was reached. For most students in the country in the late 1960s, the South was the least desirable region. Students in the South, however, didn't see it that way at all. They had a highly differentiated view of the South. Students at The University of Alabama, for example, liked their home state (i.e., they displayed a local dome effect) but rather disliked Mississippi right next door.

Preferences do change over time. Perhaps because of the popularity of skiing and the song by the late John Denver, Colorado has now replaced California as the most residentially desirable state outside of the home state for most college students. The South has changed dramatically. Peninsular Florida was always ranked higher than the remainder of the Southern states, but now states along the Atlantic seaboard (especially Virginia and North Carolina) and the state of Georgia have improved tremendously in their perceived residential desirability. These expressed preferences are being mirrored as well by patterns of net in-migration to these formerly despised states.

The biggest losers are states in the agricultural Middle West. Iowa is now last in the residential preference of Pennsylvania students for example. Why do you think this is the case? We would like you to focus your attention on two of the most maligned states— *Iowa* and *Ohio*—two states that have slipped quite a bit in the residential preferences of students at The Pennsylvania State University at least.

In order to make a fair comparison, we would like you to focus your attention on comparable Web Links. Click *Find* and pick each state in turn. Under *Web Links* choose the official site for state tourism information which is available for both states.

1. According to *Iowa's Division of Tourism* web site what is the new motto for the state?

2. Does it make you smile?

3. Can you name two places in Iowa where you can find Old World charm?

Now click on *Travel Areas* within the *Division of Tourism's* web site. A map will be displayed that is subdivided into several pieces. *Click* on the *eastern section* and answer the following questions:

4. Name two areas in this region where one could find riverboat casino gambling.

5. What unusual museum is located in Dyersville?

Now *click* on the *Southeast* corner of the state and answer the following questions:

6. What is the most crooked street in the world (located in Burlington) called?

7. Which Civil War battle was fought in this area?

Now *click* on the *Southwest* corner of the state and answer the following:

8. What three early groups traveled through this area?

9. What big band leader was born in this area of the state?

10. Where can you catch rodeo fever?

Now *click* onto the *Northwest* corner of the state and answer the following:

11. Where is a jewel-encrusted grotto located?

12. This region is called the Great Lakes resort region. What are two of the Great Lakes (HINT: You may need to go back to the Iowa map and expand the scale to find the names)?

Finally, *click* on the *Northeast* corner of the state and answer the following questions:

13. Where might one find some prehistoric burial sites?

14. The town of Decorah has a museum dedicated to immigrants of what ancestery?

Return to the home page of Iowa's *Department of Tourism* and *click* on *Iowa Facts*.

15. What is unusual about the SAT and ACT scores of students in Iowa's secondary schools?

16. How does Iowa rank in production of "the other white meat" (i.e., pork)?

Now *Find Ohio* and click on the *Web Links* that will take you to *Ohio Tourism*. We can't visit everywhere in the state, so let's focus on one of more maligned areas of the state—the area along the shoreline of Lake Erie. Cleveland has even referred to in a derogatory way as the "Mistake by the Lake".

Click in turn the information for tourist attractions number 1 (*Cedar Point*), 5 (*Rock and Roll Hall of Fame and Museum*) and 10 (*Lake Erie Islands*). Can you answer the following questions:

17. What is nicknamed "America's Rollercoast?"

18. What appears to be the most prominent architectural feature of the Rock and Roll Hall of Fame?

19. What crop was grown on the islands in Lake Erie in the 1860s?

20. What is being memorialized at the Glacial Grooves State Memorial?

21. Do you find Iowa and Ohio more attractive now?

***Keywords*:** **mental map, local dome effect, national trend, Iowa, Ohio**

de Blij, *Human Geography: Culture, Society and Space*

Part Two: Population and Space
You Are My Density

It is clear from the material that you've read in the textbook that the distribution of population around the planet is uneven. There are huge areas of population void where very few people live. Alternatively there are rich and fertile areas where the population crams onto every available parcel of arable land (i.e., land capable of producing a crop). Over half of the world's population lives on the flat and fertile floodplains of major rivers especially in South, Southeast and East Asia. These are sensitive, low-lying ecological zones subject to periodic flooding and other natural disasters. Eventually many may be at risk due to ocean level rise if global warming continues unabated.

Can you name a few of the major population void areas? They would include much of the polar region where the growing season (if there even is one) is too short to grow crops. There are also large tracts of desert in northern Africa, southwestern Africa, northern Chile, in Mexico and the southwestern United States, the outback of Australia and in central Asia and elsewhere in which it is difficult to sustain life. And finally, there are tropical rainforest areas that despite the verdant look of the landscape are incapable of supporting large populations. Tropical soils are often of poor quality because they are deficient in organic matter (humus) and leached of important (but water-soluble) minerals.

What is the consequence of large areas of population void and concentration for the countries (i.e., states) of the world? Let's focus more closely on one of the countries listed in Table 6-1 (*The Netherlands*) and another that is not (*Brazil*). That table presents you with the difference between the *arithmetic density* (i.e., population divided by total land area) and the *physiological density* (i.e., population divided by total arable land area). In some cases, there are huge differences between the two figures. For example, over 90 percent of Egypt's rapidly growing population lives within the floodplain of the Nile River or one of its tributaries. Most of the rest of the country is a sandy desert. But, the country of Egypt is very large and if we simply divide the population of Egypt (about 62 million people according to Table 6-1 in your textbook) by the size of the country, we obtain the figure that most people think of when they hear the word 'density'. It is more properly called arithmetic density.

1. What is misleading about arithmetic density in the case of Egypt?

2. What is Egypt's physiological density?

Among the countries shown in Table 6-1, Egypt displays the greatest disparity between arithmetic and physiological density.

3. Can you think of another country in which the physiological density might be many times higher than the arithmetic density?

The problem with Table 6-1 is that results are already calculated for you. That's the easy way! We want you to realize how the figures you see there were actually derived in a hands-on manner. We hope that by doing so you won't take for granted the statistics that you see in a table or even the definition of terminology such as the term 'arable'. As defined in Resource D Glossary (p. R-27 in the textbook), arable literally means cultivable. "Land fit for cultivation by one farming method or another".

4. Is land devoted to the pasturing of livestock considered arable? Why or why not?

5. What about idle land that might be brought into production but is not currently being used. Is it arable?

6. The operative word in the definition of arable is 'cultivatable'. In your own words, can you explain a possible difference between cultivated and cultivable?

Let's start with a country, The Netherlands, for which densities have already been calculated. We cannot tell directly from Table 6-1 what the amount of arable land per person is in the Netherlands, but we could derive the figure indirectly. The physiological density of the Netherlands is 4,425 persons /sq. mi., approximately 4.54 times greater than the arithmetic density of 974 persons/sq. mi. Thus, dividing the total land area by this 4.54 factor, will yield the approximate amount of land that is considered arable in The Netherlands—about 3.2 thousand sq. mi.; almost 20 percent of the total land area. Let's see if we can verify the amount of arable land using Encarta's Virtual Globe. Click on *Find.* Then click on *Country* to find the *Netherlands*. Do a search of basic statistics to see if arable land is easily obtainable. Check *Facts and Figures* and then use the *Features* pull-down menu at the top of the screen for *Statistics Center* (look especially at the *Agriculture* and *Population* categories). Interestingly, arable land per capita is not listed *per se,* but a statistic called *Agricultural Population Density* measured in persons per square kilometer is. Is this statistic really measuring the same thing as arable land per capita. Let's see.

7. What is the amount of arable land per 1,000 persons in the Netherlands?

8. What is the population of the Netherlands (use the year 2000 forecasted figure)

9. How many total hectares of arable land are there in the Netherlands? (HINT: Multiply the population (in thousands) by the number of arable hectares per thousand persons).

10. How many square kilometers of land in the Netherlands can be considered arable? (HINT: 100 hectares equals one square kilometer.)

11. What then is your measure of population per square kilometer of arable land (i.e., the physiological density) in the Netherlands?

12. How does this figure compare with the statistic called Agricultural Population Density?

13. Can you think of some factors that might account for the reason that the two figures are not exactly the same?

14. Approximately what percentage of the land area in the Netherlands is arable (HINT: the total land area in square kilometers is found is the sidebar entitled *Facts and Figures*)?

Now click on *Web Links* and open to the address labeled *Key Figures and Statistics—The Netherlands*. One can certainly find a gold mine of information at this web site. For example, click on *Agriculture, forestry and fisheries*. Once there, click on *Arable Crops*.

15. How many hectares of land are devoted to the cultivation of Triticale?

16. Answer one of the following questions: a) What is tritcale? or b) Does the Netherlands have a problem with Tribbles? Allusions to the classic Star Trek episode ("The Trouble with Tribbles") aside, it isn't easy to find the amount of arable land.

Is it any easier to determine the amount of arable land for a country not listed in Table 6-1? Without a consistent data source such as the *United Nations Food and Agriculture Yearbook,* it isn't easy to compare statistics for different countries. Click on the *Find* command and then type in *'Brazil.'* Study the map of Brazil carefully. You might note that there are few large cities in the states of Rondonia, Amazonas and Mato Grosso except for Manaus. Even Manaus was a much more important center in the past than it is at present. Why are these three states seemingly so sparsely populated? Click onto the *Web Links* and go to the source called *Brazil Quick Facts*. Note that the percentage of arable land in Brazil is estimated to be seven percent of the total.

17. Using that percentage, calculate the arithmetic and physiological densities of the country. (HINT: Click on *Facts and Figures* and you will find that Brazil is 3,286,490 sq. mi. in size, only slightly smaller than the United States and that there were 161,790,000 people living there in 1995. Now calculate the two densities (in persons per square mile.)

Keywords: **Brazil, Netherlands, arithmetic density, physiological density**

35

Part Three: Streams of Human Mobility

Like Salmon Swimming Upstream?

From at least the time of Samuel Ravenstein over a century ago, the metaphor of migration "waves" forming "streams" of migrants has been used to describe a restless world population on the move. Ravenstein is credited as the first social scientist who studied human migration in a systematic and conceptual way. One of his "laws" of human migration sounds a little like Newton's second law of thermodynamics—for every stream of migration there will eventually be a counterstream. Is this true even for the "flood" of African-Americans leaving the rural South after the turn of this century in search a better life in the Northeast, Middle West and West?

The answer is decidedly 'yes'. In fact, since about 1970, the "Great Migration", as it is often called, has reversed itself. In the 1980 census period, the percentage of African-Americans living in the South began to increase, the first time that had happened since before the turn of the century. And, the 1980 census figures were no flukes. The 1990 census continues to bear out the fact that there is now a counterstream of African-Americans moving to the South. In some cases, they are moving back to the South they had left behind decades before. In other cases, northern urban African-Americans are moving to the South for the first time. These latter African-Americans, with no particular close family ties to the South, are overwhelmingly moving to the cities of the South whereas African-Americans with family ties to certain parts of the South were often moving back "home", even to some very rural and poverty-stricken areas.

What is the impact of the Great Migration on the spatial distribution of African-American population in the United States? The statistics reveal a tremendous impact. In 1900, at the start of the Great Migration, 90 percent of all African-Americans lived in the South and the vast majority of them in the rural South where they were, for the most part, sharecroppers and agricultural workers. By 1970, at the end of the Great Migration, only 50 percent of the African-American population remained in the South and even in this home (i.e., origin) region, many African-Americans had moved to the towns and cities of the region rather than remaining on the land in rural areas. Examine Figure 11-7 (p. 128) in the textbook closely. The darkish red areas delimit counties that contain the highest relative proportion of African-Americans. How would you describe the geographic distribution displayed here? In what ways is such a map misleading? Even if only 20 percent of New York City's population is African-American, the absolute number of African-Americans would still be several hundred fold greater than the 80-90 percent figure found in many Southern counties.

Many of the counties with the highest relative percentages of African-Americans are rural counties on the Carolina-Georgia Piedmont, the "Black Belt" crescent-shaped region of rich black marl soils straddling Alabama and Mississippi and the lower Mississippi River valley from west Tennessee into northern Louisiana. What do these areas have in common? If a map of cotton production for the early 1900s were shown, the two patterns would look remarkably similar.

Why did African-Americans leave this Southern region in such large numbers after the turn of the Century? As with any migration movement that is voluntary, there are myriad causes. For one thing, the economic structure of the region was changing. Agriculture was becoming more mechanized and the need for unskilled labor to sow, cultivate and harvest the crops was diminishing. The mood of the South at the time was one of racial intolerance with public facilities that were separate but very seldom equal. There were Jim Crow laws on the books specifying where and when African-Americans could gather and where they could not. There was the ever present threat of mob lynching of African-Americans and an "invisible empire" of Ku Klux Klan chapters and organizations of their ilk intent on preserving the white man's privilege (i.e., the status quo) at all cost. Demographers, scientists who study population issues, would call all the above forces 'push factors'. There were also 'pull factors' luring rural African-Americans to the high-paying factory jobs of the Northeast, Middle West (i.e., North Central) and West regions of the country. The migration streams were boosted during WWI when the need for factory workers became critical and African-Americans provided the necessary labor. That demand continued during the rapidly expanding "roaring '20s" and then bottomed out during the Great Depression of the 1930s only to come roaring back with the war effort during WWII and the post-war boom economy of the 1950s and early 1960s.

The streams of migration are quite predictable and rational. African-American that initially resided along the Atlantic coastal plain migrated to cities in the Northeast and more of them ended up in the southernmost cities in that region (e.g., Washington, D.C. and Baltimore) than their more northerly counterparts (e.g., Boston and New York). Likewise, African-Americans from states in the East South Central region such as Mississippi were more likely to go to industrial centers of the Middle West such as Chicago and Detroit. Likewise, those originally in the West South Central region such as Texas were more likely to migrate to the Far West region, especially Los Angeles and the San Francisco Bay area.

Not surprisingly, the counterstreams of African-Americans moving to the South reflect the same pattern in reverse. It's much more likely for a retired African-American factory worker from Chicago to migrate to the Yazoo Delta of Mississippi than to, say southern Georgia or the North Carolina coastal plain.

Let's examine one city that was heavily influenced by the influx of African-Americans during the Great Migration. Click on the *Find* command and type in *United States*. Then click on *Sights and Sounds* and note the caption and the sound clip for *Chicago*.

1. What type of music is featured in many popular Chicago nightclubs?

2. Koko Taylor is a blues artist of rare talent but many of the other people mentioned including Muddy Waters who sings a form of music known as the Delta blues. What distinguishes this type of music from other genres?

3. Where is the Delta exactly?

How did so many African-Americans end up in Chicago after the turn of the century and especially after WW I? Let's examine a *Web Site* that is not reached by Encarta, but rather by one of several search engines that you might have access to (e.g., Yahoo, Lycos). One of the best sites available is http://lcweb.loc.gov. You will be linked to a Library of Congress web site location. *Click* on the area marked *Exhibitions*. Scroll down the list until you see *The African-American Mosaic: African-American Culture and History*. Within that exhibition is a section entitled *"Chicago: Destination for the Great Migration"* which is part of their collection *"The African-American Mosaic: A Library of Congress Resource Book for the Study of Black History and Culture."* Answer the following questions:

4. What Chicago newspaper was especially influential in attracting African-Americans to the city?

5. What was the African-American community that grew up on Chicago's South side in a era of Jim Crow laws and housing discrimination called?

6. Was there any geographical sorting of the African-American population noticeable within this community?

7. What was the basis of that sorting?

8. Where did the more elite African-Americans live within Chicago?

Keywords: **migration stream, counterstream, push factors, pull factors, African-Americans, Great Migration, Chicago**

Part Four: Patterns of Nutrition and Health
Not So Great (Life) Expectations

Why is it that the average life expectancy (averaging both sexes together) is greater than 80 years in the tiny principality of San Marino, an enclave within Italy and less than 40 years in the west African country of Sierra Leone? What are the determinants of life expectancy?

There is no simple answer to that question but we can do some investigating on our own using the Statistics portion of *Encarta's Virtual Globe '99*. First click on *Options* and then on *World Statistics*. The amount of information is a tad overwhelming at first. While not every demographic statistic is listed here, five variables listed under the *Health and Education* section will suffice for our purposes. The variable that we wish to explain (referred to as the dependent variable in statistics) is *Life Expectancy at Birth (1997)*. For now, we will not concern ourselves with possible gender differences such as differential access to proper nutritional and health care facilities. Using the *Statistical Tables* option, display these national level data in an array from high to low. Choose the countries in the world with life expectancies greater than or equal to 78 years. List these countries in the array.

1. How would you characterize the location of these countries if you had to generalize about them?

Now list the bottom of the array—the countries in the world with average life expectancies less than or equal to 46 years.

2. How would you characterize their pattern?

Now, let's examine some variables that might be related to life expectancy that we might posit as explanatory or independent variables. There are five in the *Health* group that should be related to our dependent variable —total life expectancy. These independent variables would include: *access to safe water in rural areas, access to safe water in the country as a whole, access to sanitation in the country as a whole, the number of people for every available hospital bed,* and the *infant morality rate*.

Let's hypothesize what we'd expect to find when comparing each independent variable separately with our dependent variable—life expectancy. One of the things that many of us in the United States take for granted is access to safe drinking water but such is not the case in many parts of the Third World. Contaminated water can cause all sorts of problems from the spread of deadly diseases like cholera to the more mundane but equally deadly diarrheal diseases like dysentery. Such water-borne diseases kill many young infants and as such, safe water and infant morality rates are undoubtedly interrelated.

Access to safe water is sometimes easier to obtain in the city where there are water purification plants than in the rural countryside where the water might be drawn more directly from contaminated rivers or wells. Access to sanitation in a country should also cut down on the number of diseases spread by contact with contaminated water sources as well as hookworm and other diseases that can be obtained from contact with human and animal wastes.

The number of people per available hospital bed would be a measure of the quality of the health care delivery system. Advanced economies should have lower numbers than their Third World counterparts.

Finally, a large contributor to a relatively short life expectancy is the infant mortality rate. That is, the number of children born alive who make it through the first year of life. Lack of access to health care facilities or a safe water supply will certain drive up the infant mortality rate.

Using the *Table of Statistics* option, compare the arrayed data for access to safe water (total) with that for the dependent variable.

3. Are there any "overlaps" (i.e., countries that appear on both lists)?

Six of the twelve countries at the high end of the life expectancy table also have the highest value population with access to safe water (100 percent). Alternatively, four of the 10 countries at the low end of the water accessibility scale (33 percent or less of the population possessing such access) are among the countries with the lowest life expectancies. The results are similar if we consider just access of rural areas to safe water supplies. Again, six of the countries in which 100 percent of the rural population has access to safe water are also those countries with the highest life expectancy. Likewise, three of the 10 countries with the worst rural access to safe water (less than 17 percent) have the lowest life expectancies. Access to sanitation is a little less clear-cut although two of the 10 countries with the worst access to sanitation were among the 10 countries with the lowest life expectancies.

4. Why are there fewer matches between dependent and independent variables in the case of the number of persons per hospital bed?

5. Two of the countries with the lowest life expectancies also had the highest number of persons per available hospital bed. Which countries are they?

6. At the other end of the array showing the countries with the best access to hospitals as measured by the number of persons per available hospital bed there was only one match. One might conclude that this surrogate (i.e., easily measured substitute) of access to health care delivery is not the best. What is unusual about the countries that have the best access to hospital facilities?

All but one is an independent state that emerged from the breakup of the Soviet Union. They have hospitals, but are those facilities equipped with the latest technologies and medicines? That is highly doubtful.

The best match is the relationship between infant mortality rate and life expectancy. Eight of the ten countries in the world with the highest infant mortality rates (mainly confined to the continent of Africa) are also those with the lowest life expectancy. Likewise, six of the twelve countries with the lowest infant mortality rates (less that eight infants dying for every 1,000 born alive) also have the highest life expectancies.

7. Which countries are they?

8. Why, in your opinion, isn't the United States among those countries of the world with the lowest infant mortality rates?

Keywords: infant mortality rate, life expectancy, access to health care and sanitation, dependent variable, independent variable

de Blij, *Human Geography: Culture, Society and Space*
Part Five: Geography and Inequality
A Woman's Place is in the House...and in the Senate

This title appears on many car bumpers and it turns an old cliché on its head. Women are working outside the home in record numbers now and, as they assume greater positions of responsibility, the disparity in wage level between what a man is paid and what a woman is paid for a comparable job is closing. But, a disparity is still present and many women feel they are called upon to balance work and home life to a much greater extent than their male counterparts. Likewise, professional women often complain of a "glass ceiling", an invisible but nonetheless very real barrier to their progress through the professional ranks of major businesses simply because of their gender.

These are serious disparities to deal with in our modern society. But, there are even greater disparities between men and women in other parts of the world. At worst, women are treated like chattel, as objects of value incapable of making their own decisions outside the confines of the home of the girl's father or woman's husband. In some patriarchal societies women are denied access to even the most basic of opportunities based solely on their gender.

Let's examine two statistics that might indicate the lower status of women in many traditionally based societies—female literacy and the fertility rate. Click on *Features* and go to *Statistics Center*. Under the heading of *Education* is the statistic *Literacy Rate, Female*. Interestingly, almost half of the countries in the world, including the United States, do not provide these data. Examine the data in tabular form arrayed in order of the percentage of females that are literate.

1. What can be said about the countries at the bottom?

2. Where are they located?

Now, examine the same statistic for males (i.e., *Literacy rate, male*). Again, array the data in descending order of percentage of the male population that is literate.

3. Is there a strong positive association between countries where females are mostly illiterate and those where males are also found in the same basic condition?

The answer is decidedly 'yes', but even in these countries, males typically have a higher literacy rate.

4. What is the approximate factor (i.e., multiple) that would make male and female literacy in these countries approximately equal?

42

Another sobering statistic is the *fertility rate*. It is defined as the number of children that the average woman gives birth to during her childbearing years (usually defined as age 15-45). While still in *Statistics Center* click on *Population* to find this statistic. Take a minute to comprehend that in the countries at the top of the array, located mostly in northern and sub-Saharan Africa, the average woman gives birth to seven or more children during her child-bearing years! Is it little wonder that women in these countries seldom work outside the home?

Let's go beyond the statistics. Click on the *Find* command and then type in *'women'* as the content subject area. An overwhelming number of sites come up, but we want you to confine your attention to those that first appear on the screen without moving the side bar. That will be sufficient. If you want to explore the subject of women in various societies further, you might like to examine the other sites on your own.

Focus on the role of women in *Saudi Arabian* society and then answer the following questions:

5. How many wives can a Saudi man have under Islamic law?

6. Does he need to seek the permission of his other wives before he can take another?

7. What is the veil called that women in Saudi Arabia wear that covers their face? Can a woman ride a bicycle in public?

8. Can an unmarried man and woman go on a date alone in Saudi Arabia?

9. What do they call the money that a man may have to pay the family of a bride in order to marry her?

10. Can a woman doctor in Saudi Arabia examine a man?

11. Under what conditions is that possible?

Religion may have an effect on the birth and fertility rates. The Roman Catholic Church, for example, forbids the use of artificial means of contraception among its adherents. This conservative view was recently reinforced in *Veritas Splendor*, an encyclical (religious statement) issued by Pope John Paul II.

12. Why then do you think the nominally Catholic countries of Italy and Spain now have among the lowest fertility rates in the world?

13. Could it have something to do with the percentage of women in the labor force (i.e., women working outside the home)?

One might hypothesize that if a lot of women worked outside the home, their desire and ability to raise many children would be diminished. From the *Features* menu at the top of the screen choose *Statistics Center*. Then choose the *Economics* category and the variable called *Laborforce, female participation as share of total (percent of total labor force)*.

14. How do female labor force participation rates in Italy and Spain compare with most other developed countries in North America and Europe?

15. How do Italy and Spain compare with most of the non-Islamic third world countries in Asia?

Keywords: **Literacy Rate, Fertility Rate, Women's Issues, Gender**

Part Six Landscape and the Geography of Culture

What'll it be: Lethargic and Clever or Vigorous and Stupid?

One of the unfortunate legacies of early theorizing in geography derives from a school of thought known as **environmental determinism**. That school goes back to the ancient Greeks who first coined the word 'geography' to define their scientific attempt to describe and understand the known world of their time. Interestingly, one of the contributors to this body of Greek literature about geography was Hippocrates, the great physician. Doctors to this day still pledge allegiance to the Hippocratic oath. They promise to "first, do no harm". Hippocrates wrote *On Airs, Waters and Places*, an early geographic compendium in the 5th century BCE. In it, he tried to explain the progress of human civilizations around the known world of the time—portions of Europe, Asia and Africa. He noted that within the frigid climate of the northern European plain were roving bands of kinship-based societies that we know today as Huns, Visagoths and Ostragoths. These tribes were not as advanced as the Greeks in many of the finer aspects of civilized culture (e.g., theatre, the arts, athletic competition, scholarly pursuits). But, these tribes were a strong, militaristic lot. It was as if, being too far from the sun's warming rays, these people lacked intelligence but made up for it by their physical prowess.

Likewise, to the south of Greece in the "torrid" zone closer to the equator lived groups of people who were very intelligent and clever, having sailed their dhows around the Red Sea and beyond to uncover the riches of sub-Saharan Africa. But, these people were judged to be living too close to the sun and were, therefore, lethargic. They had to take a break during the heat of the day for example as if the sun had sapped them of their energy.

In between, was the temperate (actually Mediterranean) climate of the Greek peninsula that, like Goldilocks inspecting Baby Bear's things, was found to be just right. Not too hot. Not too cold. Just the right combination of vigor and intelligence. In other words, Hippocrates was not only **ethnocentric** but he explained the differences in the degree of sophistication of the world's known civilizations by their location relative to aspects of the physical environment—in this case, climate. That is, the physical environment was held to determine people's behavior. That, in a nutshell, is what environmental determinism is all about.

Your textbook presents a more modern day application of the same theoretical stance—Ellsworth Huntington's view of climate and civilization (see Figure 19-1 on p. 230). Writing in the early part of this century, Huntington, a geography professor at Yale, was very influential. His most famous "theory" may have been **the beating heart of Asia** in which he argued that the "ruthless, ideal-less horsemen" that we call the Tartars or Mongols have influenced the histories of both **China** and **Russia**. Like a beating heart, when the climate cycle was in a period of adequate rainfall, the range of the nomadic Mongols and their herds of horses, camels, yaks and other livestock, **contracted** to central Asia and present-day **Mongolia**. But, when the climate went into an extended period of drought, the range of forage, of necessity, had to **expand**. The ancient Russians eventually had to move their capital from Kiev in the grassland prairie to Moscow in the

45

forested zone so that they could better defend themselves against these mounted invaders. Click on *Find* and then **Russia** and choose the **Physical Features** type of map.

1. What two physical features did the mounted horsemen from Mongolia encounter as they rode toward **Kiev** (the present day capital of the **Ukraine**)?

Now switch to the **Climate** map of the same area.

2. In what climatic regime do we find Kiev?

3. In what climatic regime do we find Moscow?

How did the economy of China differ from that in Mongolia at the time when Huntington argued his theory was appropriate? The Chinese were, for the most part, farmers and not nomadic herdsmen. The richest farmland was along the floodplains of the two major rivers that run through China. Click on **China** and then choose the *Physical Features* type of map.

4. What are the names of those rivers?

5. Where is the mouth of each of these major rivers?

What did the Chinese do to keep the Mongols out of their agricultural villages located along the floodplains of the two great Rivers? Click on *Find* and then *Content* and type in **"Great Wall of China'** The Great Wall of China is one of the few man-made features that can be seen by the astronauts from space.

6. When was this most massive of man-made features begun?

7. When was it completed?

Geographers suffered a "burnt fingers" reaction to environmental determinism. They dabbled in theory and were eventually excoriated for it. Many geographers unfortunately clung tenaciously to this model long after it was rejected and repudiated in most other disciplines. This way of thinking—that there is a one-way causative relationship between the physical environment and human behavior—is too simplistic. Some geographers overreacted however and denied that the physical environment had any effect on human behavior. Behavior, it was thought by these geographers, is constrained only by the human mind itself.

8. Do you think that this more humanistic way of viewing the nature of human-environment relationships is an example of "throwing out the baby with the bath water?" Why or why not?

Keywords: **Environmental Determinism, Ellsworth Huntington, "beating heart of Asia", Hippocrates**

de Blij, *Human Geography: Culture, Society and Space*

Part Seven: Pattern of Language

You Are What You Speak:
The Rhythms of African Lingua Francae

The intricate mosaic of African languages reveals much of the continent's complex blend of indigenous and colonial history and culture. No less than seven language families are represented ranging from India-European in South Africa, Malay-Polynesian in Madagascar, to Afro-Asiatic in the Saharan north. Along the border between the Afro-Asiatic languages of the north and the Niger-Congo languages of sub-Saharan Africa lies a transition zone where two major lingua-francae, Swahili and Hausa, dominate. Swahili evolved from Bantu, Arabic, and Persian languages while Hausa is a blend of Sudanic and Afro-Asiatic tongues. Lingua-franca literally means "Frankish language" as explained in the glossary of your textbook. This is a language that can be more generally understood across a wide geographic region than any one of the several of the component languages. But language is more than words and phrases and literally shapes our world (cosmology). Language can tell a story within a story. It can also be a song and Africa is a wonderful place for listening.

PART I: Dancing Across East Africa

For this activity will need to use Encarta Virtual Globe 99 to visit three countries, Somalia, Tanzania and Cuba. First go to **Somalia** and in the *Content* box click on **"Somalia: Urban Love Song,"** and **"Somalia: Trance Music,"** via *Sights and Sounds*. View the slide, read the captions, and listen to the music sample from the coastal town of Baraawe.

1. How did colonialism and immigration affect the music styles of Somalia?

2. What are two examples of how Arabic culture has influences Somalian music?

3. What is the purpose behind the "Possession Dance" and trance music?

Now go to **Tanzania** and using the *Context* box, select **"Tanzania: Urban Dance Music"** to view a slide of Dar es Salaam and listen to the sound clip. To complete this part of the activity you will also need to take a side trip to **Cuba** and select *Sights and Sounds* to listen to **"Cuba: Popular Dance Music."**

4. Does Tanzanian urban music sound similar or different from that of Somalia and why do you think this is the case?

5. How has the Tanzanian government influenced musical styles in the country?

6. How do you suppose Tanzanian music has been influenced by Cuban dance music and conversely, how has African music helped shape Cuban music traditions?

PART II: Hausa The Setting Sun: Musical Traditions in West Africa

Hausa is the lingua franca for millions of people living in a cluster of countries in West Africa. For example, this is the language of commerce in Niger and the language spoken on "Radio Niger." A visit to Niger will illustrate how language serves as an integral part of culture. Using the *Find* box, create a map of **Niger** and then select *Society*. To answer the following questions, read the **"Demographics," "Language,"** and **"Religion,"** sections and then click on *Sights and Sounds* and view **"Life in a Hausa Village."** Be sure to listen to **"Hausa Praise Song,"** in and **"Niger: Fulani Music,"** *Sights and Sounds* to hear music samples.

7. How has Islam influenced Nigerien history and culture?

8. What is the official language of Niger and why do you think it is spoken only by a small minority within the country?

9. Who are the Fulani people and how would you describe their music?

10. What is the purpose of the Hausa "praise song" and how does it compare to Fulani music in sound and meaning?

11. Now that you have heard music samples from east and west Africa, describe their differences and similarities and offer some possible explanations for these contrasts and parallels.

Africa is not the only place were you find lingua francae. Using the *Find* tool, type in the words **"lingua franca"** and see what comes up in the *Content* box. Click on that single entry and answer these questions:

12. What is the name of the place you found when you clicked on the item in the *Contents* box?

13. Why do you think this location would have a need for a lingua franca?

14. Do you believe this place has very much in common with the lingua franca examples in Africa? Why or why not?

Keywords: **Africa, language, lingua franca, Swahili, Hausa, music**

Part Eight: Geography of Religion

Christianity Worldwide: Plain Vanilla or 31 Flavors?

As you have gathered by now, we will sometimes dwell on aspects that the textbook trips over rather lightly. This is an example of complementarity. Together, the book and this companion volume can then cover a wider swath of related material. It is common for introductory cultural textbooks to focus on the exotic and the faraway, especially when it comes to the subject of religion. Not many of us know a great deal about the state religion of Japan—Shinto—for example. Textbooks will often dwell on topics such as: 1) the differences among the Eastern religions; 2) interesting cultural practices such as the avoidance of certain foods and drink within the belief systems of different religions; and, 3) the diffusion of the main branches of Islam from their core area in southwestern Asia.

Your textbook authors note that there are three truly universal (and universalizing) religions in the world—Islam, certain forms of Buddhism and Christianity. The one that usually gets short shrift is Christianity probably because it is too close to home, too familiar and seemingly too well known. But is Christianity really practiced similarly throughout the world or it is reflective of the local indigenous culture in which it is found? Let's see what we can find out. The United States presents an interesting mosaic of practice in an ostensibly secular but majority Christian country. Take a look at the map of dominant religions in the United States (Figure 25-2 p. 315). Whether it is the Mormon in Utah or the French Catholic in Louisiana, all fall into the general category of Christian. And although the Jewish population of the greater New York metropolitan area supposedly exceeds that of Israel, Judaism is still not so dominant that it could be detected on a map of this scale. The regional distribution of religion in the United States is often reflective of the origins of the earliest settlers in the region. For example, where did most of the settlers of the upper Midwest (e.g., Minnesota, the Dakotas) come from originally? What religion is the state church in those countries (if any)? Let's now look at a few countries in Europe. What about the birthplace of the Protestant Reformation—**GERMANY**? Click onto the *Find* command and then onto **Countries**. When you have Germany specified, click on the sidebar where it says **Society**. Within that choose **Religion**.

1. Where are most of the Lutherans in Germany located?

2. What about the Roman Catholics—where are they located?

3. The article mentions a growing Muslim presence in Germany. What might be the source of Islamic **immigration** into Germany or do you think this increase represents instead a **conversion** of an increasing proportion of this large secular society to the tenants of the Islamic faith?

Let's examine three more countries in Europe—**Belgium**, with two distinct culture realms; **Poland**, homeland of the current Pope of the Roman Catholic Church; and **Bulgaria**, a former Warsaw-pact nation in Eastern Europe. Bulgaria was formerly part of the Islamic Ottoman Empire, but a country that, like Poland, also endorsed the official state atheism of a Communist regime until very recently. Click onto the *Society* sidebar and then choose the *Religion* category for each of the three European countries in turn.

4.   If there is such a linguistic and cultural division between Flemish northern Belgium and the Walloon population of southern Belgium, why is such a large percentage of the population in both parts Roman Catholic?

5.   The write-up suggests that Flemish residents are "more religious" than the Walloons. What do you think that means?

6.   If the Flemish area in the north has a closer affinity to the Dutch and the Walloons in the south to the French, why is Catholicism seemingly stronger in the northern part of the country, adjacent as it is to the influence of the Reformed Church (Protestant) in the Netherlands?

Poland is a very Roman Catholic country.

7.   When did this conversion to Catholicism take place?

8.   What town in Poland is Pope John Paul II from originally?

9.   What role did the Roman Catholic Church play in the opposition to Communism?

10.   In your own words what does it mean to coexist but not cooperate with the Communists?

11.   What is unusual about the Uniate faith to which a minority of the Polish population adheres?

Click on the photograph of the **Alexander Nevsky Cathedral** in Bulgaria and read the caption.

12.   Where is this cathedral located?

13. What does the construction of the cathedral symbolize?

14. Why does Bulgaria feel indebted to Russia?

15. Why are Bulgarians members of the Bulgarian Orthodox Church rather than being adherents of Roman Catholicism?

16. Can you briefly describe The Great Schism and the division between the Roman and Byzantine Empires?

Keywords: **Roman Catholic Church, Protestantism, Orthodox Church, Belgium, Poland, Bulgaria**

Part Nine: Cultural Landscapes of Farming

Let Them Eat Jute: The Colonial Legacy in World Agriculture

In most Third World countries, it is all the farmers can do to keep up with the internal demands of their country for foodstuffs. Very little surplus is produced for export. In fact, many of these countries have to import foodstuffs despite the high proportion of the labor force engaged in agriculture. The textbook discusses the Third or Green Revolution in which high yielding varieties of grain have helped to stave off world hunger by increasing the world's food supply at least temporarily.

Long before this Third Revolution began, however, valuable arable land that could have been devoted to food crops was being taken out of production. Some of this land, especially in the tropics, was devoted to the growth of crops that could not be grown in Europe (e.g., tea, coffee, tobacco, bananas) but for which there was a growing demand. Much of this land was organized into plantations in which scientific principles of horticulture were applied on a large, commercial scale. By buying necessary inputs in bulk and marketing their products to a mass market, these plantations often achieved economies of scale compared to the small plot cultivators they might have been competing against.

In addition to "feeding" the culinary desires of their colonial overseers, commercial crops were also grown that could not be eaten. These would include, but not be limited to, indigo, cotton, jute, sisal, and rubber. Many of these fiber and industrial crops could only be grown in tropical or subtropical countries located throughout Asia, Africa and South America.

Let's look at two such industrial crops—cotton and jute. Everyone knows what cotton is, but what is jute and what is it used for? Click on *Find* and then *Content* and type in the word 'jute'. Several sites will come up. Click on the slide called **Sorting Jute for Export in Bangladesh**. As you can see, jute is a cane-like plant with strong, stringy fibers that can be turned into cordage and sacking material. What we call "gunny sacks" are made from jute. Often the backing of tufted carpets is made of jute. It is a very tough natural fiber.

1. According to the caption, what has been the effect of synthetic substitutes on the market for jute in the extremely poor country of Bangladesh?

Now click on **the IJO (International Jute Organization).** It is clear that the major growers of jute are trying to form a cartel-like arrangement with the major purchasers of the fiber.

2. What are the five countries that are exporters of the product?

3. How would you characterize the 21 countries that have signed the agreement as importers?

Now, click on *Features*, then *Statistics Center* and then *Agriculture*. One of the variables that is mapped at the world scale (and listed in tabular form) is **Crops, cotton production (metric tons)**. Countries with large populations grow cotton (e.g., India, China, Pakistan, United States) but also cotton is relatively more important to some smaller countries for which this industrial crop represents a major source of export income.

4. What are the top ten cotton producing countries in the world?

Let's focus on some of these countries and others that rank high in production through *Sights and Sounds*. Click on *Find*, and then *Content* and type in 'cotton'. Many of the entries on the screen (without moving the sidebar down) are pictures of cotton production in some of these countries. Click on the slide **Cotton Harvesters near Samarqand** in Uzbekistan. As the caption notes, cotton is that country's leading source of export income, but it has come into production at a terrible cost.

5. What are some of the ecological and social problems caused by the commercial production of cotton in that country?

Now look at **Sewing Egyptian Cotton.**

6. What makes **EGYPT'S** cotton so special?

7. Why do you think that long staple cotton (i.e., long fibers) is so in demand for clothing and textiles relative to its shorter staple counterpart?

Now click on **Valuable Cotton Crop for Azerbaijan.** The caption of this slide focuses on changes in cotton production that have taken place since the fall of the Soviet regime.

8. Why haven't state farm collectives been done away with completely?

Now click on the slide of **Transporting Cotton in Burundi.** Compare and contrast the method of transportation cotton in this central African nation with that used by Azeri farmers in the previous slide.

9. Which method is more likely to accrue economies of scale?

Finally, click on **Tajik Cotton.** Tajikistan is another of those former Soviet republics in central Asia.

10. Why is irrigation of cotton there so disastrous?

53

Click on a map of this central Asian region showing the independent states of Kazakhstan, Uzbekistan, and Tajikistan.

11. What body of water might form the most important source of irrigation for cotton crops in these newly independent countries?

This body of water has shrunk to about half of its former size in the last forty years—an ecological disaster of the worst kind for all of the surrounding states.

Keywords: **Cotton, Jute, Bangladesh, Egypt, Uzbekistan, Azerbaijan, Tajikistan**

Part Ten: The Urbanizing World

Planet of the Apes? The Primate City Distribution

Excuse the play on words. The term 'Primate' in the chapter title has nothing to do with the apes you might find in the primate house of your local zoo. Rather, the term is related to the concept of primacy or dominance. Urban geographers have been curious about the role of cities as the engines of state economies since at least the time of Mark Jefferson, the geographer who, in the early part of this century, first coined the term 'primate city distribution.' What is a primate city distribution exactly and what is the alternative? According to a recent textbook, a primate city is defined as "an urban center more than twice the size of the next largest city in the country; has a high proportion of its nation's economic activity; most obvious in the less developed world." (Fisher, ed. 1992, p. 703).

The alternative to the primate city distribution is called the rank-size regularity or rank-size rule. If the cities in a country conform to the rank-size regularity, one can easily estimate the population of a city of any given rank simply by dividing the population of the largest city by the rank of the city in question. For example, the fourth ranked city should be 1/4th the population of the largest center and so on. When graphed on double logarithmic paper (i.e., the logarithm of rank is displayed on the X-axis and the logarithm of city population on the Y-axis), a straight-line relationship between city population and city rank is found.

These empirical generalizations about the relative size of cities within countries continues to fascinate geographers despite the fact that neither the primate city distribution nor the rank-size rule says anything about <u>where</u> the cities are located within the country or relative to each other.

The geographer who has gotten the most "mileage" out of these regularities is Brian J.L. Berry. More than thirty years ago, he "explained" the rank-size rule as the equivalent of a system at equilibrium. Granted, it is dynamic equilibrium. Some cities can grow meteorically, others can almost collapse but there is enough systemic order to maintain the rank size regularity despite the self-canceling effects of rise and fall just noted. In every decennial census period from 1790 to the present, the United States urban system has manifested an approximate rank-size regularity.

A bit more controversial is the notion that primacy implies underdevelopment and rank-size distributions are symptomatic of developed economies. Berry studied 38 different countries in the world and his results were mixed. Nonetheless, he concluded that countries with the following characteristics would be more likely to conform to a rank-size regularity: 1) long history of urbanization (hence China might be rank-size even though not well developed economically when Berry was writing); 2) large size (hence Brazil with its federal system and need for regional centers to effectively integrate the country might be rank sized); 3) complex and multi-sectoral economy (hence the Czech Republic with its diversified, manufacturing economy might be more likely to be rank-sized than, say, Albania with its much simpler agriculturally-based economy; and 4) no recent history of colonization (hence the United States which has been independent since 1776 would be more likely to be rank-sized than say, Ghana, which gained its independence in the 1960s and for which Accra, the capital and major port, was the nexus of most colonial investment. Others argue that the relationship between city size distribution and level of economic development is murky at best.

We want you to decide if you think there is anything to the notion of primacy being associated with underdevelopment. Click on *Find* and then *Countries* and record three variables for each of the countries in a list that will follow. Click on the *Facts and Figures* sidebar for each country in turn. Under *Basic Facts* find the population of the largest city and the population of the second largest city. Under *Economy,* find the per capita gross domestic product (GDP) if it is available. Do this for the following ten countries—Botswana, Brazil, Canada, China, Czech Republic, France, Germany, United Kingdom, United States, and Vietnam. Five of these countries are considered developed (i.e., Canada, France, Germany, United Kingdom and the United States) and the other five are categorized as developing or emerging. Do the five developed countries all manifest a rank-size regularity?

To answer this, calculate the ratio between the first and second largest city in the country (ignoring for a moment that the figures may have been recorded at different times). If that ratio is greater than 2.0, the country is said to have a primate distribution.

1. How many of the five developed countries display primate distributions?

2. How many of the five emerging nations do?

3. How would you describe the correlation (association) between this measure of primacy and per capita GDP?

4. Do you think that Berry was justified in his characterization of the rank-size regularity as a measure of development? Why or why not?

5. Even the United States displays a value indicating the primacy of New York City. Will Los Angeles close the gap? Why or why not?

Keywords: **primate city, rank-size rule, dynamic equilibrium, primacy**

Part Eleven: Cultures, Landscapes, and Regions of Industry

Maquiladora? Is That the Latest Latin Dance Craze?

No, maquiladora is not the latest dance craze that will finally replace the macarena. It is instead a newly-created Spanish word based on the concept of the maquila, a tax paid on flour for the privilege of having the grain ground into flour at a mill. The tax was equivalent to the value added by the manufacture of the grain into a more useful and valuable product. The concept has been broadened to include all kinds of manufactured goods. It usually refers to the practice of locating twin plants, one on each side of the international boundary between the United States and Mexico. These plants are able to seize upon those factors of production in which each country has a comparative advantage. In the case of Mexico it would be a labor advantage especially if workers are paid in Mexican pesos. In the case of the United States, it would be capital availability, the distribution network and the transportation system to deliver the products to their customers.

Let's focus on one area that has been greatly affected by the maquiladora concept—the urban conglomeration of San Diego, CA and Tijuana, Mexico. Indeed, these two communities on the international border are more functionally integrated than they ever have been in the past. They say that a picture is worth a thousand words. If that is so, I would invite you to witness a visual display of La Frontera, the border between San Diego and Tijuana. To do so you will have to type in the following web address: **http://music.calarts.edu/old/AlejandroRosas/** You will be connected with a web site entitled the ***Border Project/La Frontera***. Now click on ***The Factory***. Within this website there is both important text and a series of 15 slides (***Photographic Archive***) with informative captions. After reading the material about ***The Factory*** and examining the accompanying slide set, answer the following questions.

1. How many persons were employed in Tijuana maquiladoras in 1995?

2. What rank do these maquiladora plants have as a generator of income for the Mexican economy?

3. Of the approximate 2,300 maquiladora plants in Mexico, how many are located in Tijuana?

4. When did the Mexican government build the Nueva Tijuana industrial park?

5. Why was the industrial park built?

6. What is Mesa de Otay?

7. What percentage of the maquiladoras is linked to southern California according to a 1986 survey?

57

8. What provision in the North American Free Trade Agreement (NAFTA) has aided the NAFTA nations (i.e., Canada, the United States, and Mexico) when non-North American countries wish to locate maquiladora plants of their own?

9. What Korean companies have recently built facilities in Tijuana as a result of the NAFTA?

10. What sector of Japanese manufacturing has built or plans to build assembly plants on the border?

11. What are they building on the San Diego side of the international border?

12. How many tractor-trailer trucks does Sony have going back and forth across the international boundary of the United States and Mexico on a daily basis?

13. How many Americans cross the international boundary to go to work everyday in Tijuana according to a 1994 survey?

14. How much do Mexican visitors spend on shopping goods in San Diego and its American suburbs each year?

15. Why is Otay Mesa on the San Diego side of the border finally beginning to develop?

16. According to the slide captions, what does the sign of a man, woman and child running mean?

17. Where are these signs posted (Slide 10)?

18. What, according to the caption accompanying Slide 11, might be erroneous about the sign?

19. According to the caption that accompanies Slide 14, who are the main customers for the Factory Outlet Mall in San Ysidro, CA?

One of the biggest downsides to maquiladora plant locations is their impact on the environment, especially the pollution of the already near-toxic Tijuana River and the groundwater supplies. Be sure that **'Tijuana'** is typed in the ***Content*** portion of Encarta's sidebar and then ***Click*** on the web link. ***Click*** again on a web site entitled ***Tijuana River Pollution and Maquiladoras*** for further information about this nagging problem of environmental enforcement especially on the Mexican side of the international boundary.

Keywords: **Maquiladora, Tijuana, San Diego, NAFTA**

The Yacht Sea People: Roll the Dice and End Up in Vancouver

There is a joke in British Columbia, Canada that Vancouver should be renamed Hongcouver or Vankong because of the large influx of former residents of Hong Kong into this thriving metropolitan area and Pacific Rim seaport. In fact, Vancouver has become the fastest growing metropolitan area in North America largely as a result of the recent Diaspora of former residents of Hong Kong, fleeing the country before it reverted to China in July 1997. Many of these Hong Kong residents are very wealthy and feared expropriation of their wealth by the Chinese government despite assurances that it would not happen.

Where have these Chinese people from Hong Kong moved to within the city of Vancouver? One guess would be in their traditional neighborhood often referred to as Chinatown. Despite the official change of name to the International District, locals still refer to the area as Chinatown with its distinctive red street lanterns and red telephone booths. Red is a color that brings good luck in Chinese cosmology. How large is the Chinatown of Vancouver? Click on *Find* and then *Places* and type in '**Vancouver**'. Then click on Vancouver on the map that is displayed and then click on *Sights and Sounds* and examine the caption in the slide of **Chinatown**.

1. Where is the only Chinatown in North America that is larger than the one in Vancouver?

2. When did the Chinese begin to come to Vancouver?

3. As a group, would you call them a Johnny-come-lately to the city?

The "problem" from the perspective of many long-established residents of the city is that the new Hong Kong Chinese are very wealthy and not content to confine their locational choices within the traditional areas that Asian immigrants have lived in the past. For these *nouveau riche* Chinese, nothing less than the best neighborhood in the city will do. What is that neighborhood? Click on *Find* and then *Places* and type in '**Vancouver'**. Now click on the *Web Links* and open to the web site entitled **Expedia World Guide-Vancouver**. Within that web site, there is a sidebar called **Vancouver Guidebooks** with a pull-down menu. Click on the **Neighborhoods** option and scroll down to the **West Side**. As with many American cities, Vancouver's neighborhoods are defined by occupational, demographic and income characteristics. Working class, blue collar neighborhoods are on the east side of town (e.g., East Van) while the toney and upscale neighborhoods are on the West Side. Young affluent professionals (i.e., Yuppies) might opt for a condo in **Kitsilano** for example. This is the neighborhood from which the environmental activist group Greenpeace first got its start. But, the really ritzy neighborhood is **Shaughnessy Heights**, a verdant leafy paradise of fine homes with views of the inlet and parks. Traditionally, the makeup of the neighborhood was very Anglo; many of the elderly residents tracing their ancestry many generations back to the United Kingdom. The Hong Kong Chinese that are moving into the area often tear down

59

half-timbered English Tudor style houses and replace them with bold showcase houses, the "footprint" of which takes up practically the entire lot leaving very little, if any, green space. Whereas before the house might have had 3,500 sq. ft. of living space, it's successor may be more in the 12,000 sq. ft. class. Long-time residents of the neighborhood claim they aren't being racist, but rather that these garish new upstart "monster houses" ruin the quiet, genteel look of the neighborhood. What do you think?

4.   What difference might it have made if these very wealthy Chinese had remained in the International District (i.e., Chinatown) located closer to the downtown area?

In 1975, after the United States pulled out of Vietnam, many persons fearful of retribution by the Communist North Vietnamese who took over the country, fled to other nearby countries. Many placed themselves in harm's way by plying the sometimes-treacherous waters of the South China Sea in rickety, makeshift boats. They were referred to as the "boat people" and their plight drew worldwide attention as some of their neighbors were unwilling to grant them political asylum. Was the departure of Hong Kong's "Yacht People", as they are sometimes factiously called, really necessary? Click on *Find* and then on *Countries* and type in **Hong Kong.** Look at the map of the autonomous state of Hong Kong, for the foreseeable future to treated as a Special Administrative Region (SAR) of China and granted a great deal of local autonomy. Of course China acceded to these conditions with the British in the historic accords that were worked out between the two nations. Prior to July 1997, Britain had a 99-year lease on the colony. Why do you think that China acceded and is downplaying its authority to control events in Hong Kong?

Click on the map of Hong Kong itself and go to *Web Sites* for **Hong Kong**. One is called **Hong Kong Picture Archive**. The actual web site is entitled **Hong Kong Scenic Homepage**. Go to the place where the archive is **Categorized by Year** and click on the year **1997**, the year the colony was handed over from Great Britain's control to that of China.

5.   How many photographs are there of the handover ceremony which took place on July 1, 1997?

Now click on the site entitled **About Hong Kong 1997** under the category of **Related Sites.** Scroll down the page for the news on the day of the handover (July 1, 1997—listed as 1/7/97, the British convention of the day preceding the month). Read two articles: **"Defiant touches in Patten's last goodbye"** and **"Hong Kong to be tolerant society, but with 'Chinese values' says Tung"** and answer the following questions.

6.   Who is Patten?

7.   Who is Tung?

8.   According to Patten, what precipitating event caused Great Britain to take over Hong Kong in the first place?

9. What do you think he meant when he said "democracy and human rights are not Western concepts"?

10. What are some of the traditional Chinese values to which Mr. Tung refers?

11. Is 'communal responsibility' a code word for the license to crack down on individual expressions of protest against the authorities?

Suppose for a minute that you are a wealthy Hong Kong industrialist who is still contemplating a move to Vancouver or elsewhere within the Pacific Realm.

12. After reading Mr. Tung's statements of July 1, 1997 in the South China *Morning Post*, would you be convinced that you could conduct your business as it was when Hong Kong was a British dependency? Why or why not?

13. Do you think the recent downturn in stock markets and currency devaluations throughout much of Southeast Asia has anything to do with the Chinese takeover of Hong Kong? Why or why not?

Keywords: **Vancouver, Hong Kong, overseas Chinese, "Yacht People"**

Chapter 1: True Maps, False Impressions: Making, Manipulating and Interpreting Maps

Large Map, Small Scale; Small Map, Large Scale

One of the most confusing aspects of cartography has to do with the professional's use of the phrase large (or small) scale compared with the layperson's conventional usage of the same term. To the layperson, if a map depicts a large area of the earth's surface, say an entire continent, that is a large- scale map. By the same token, a plat map of a suburban subdivision would be a small-scale map because the total area depicted is relatively small. This is exactly the opposite of the way the geographic professional uses the terms.

Let's use the tools available in Encarta to reinforce this point. When Encarta first boots up, it is set for a view of the *World*. What you see in the mapped image is a spherical representation of the earth as if viewed by a satellite or space ship hundreds of miles above the surface of the earth. We use the term representation because it is literally impossible to portray the spherical earth surface on a flat, two-dimensional map without sacrificing direction, shape or distance (or some combination of two or more of those critical aspects). The Mercator projection developed in the 16th century, for example, distorts the polarward extensions of the map such that Greenland appears as large as all of South America when in fact it is only about one-seventh its size. And yet, the Mercator projection is still very useful for navigational charts because a straight line can represent the shortest route between two points.

Let's see if map scale differences and the nature of the real shortest path between two points can be illustrated using *Encarta's Virtual Globe '99*. With the map still at the small scale (i.e., with the approximately 9-cm. long legend bar at the bottom right equaling 6,000 km.), place the mouse on the smaller map of the world in the upper left-hand corner of the image on the screen. See how adept you are at holding your finger on the mouse and dragging it on that world map until you have the "cross hairs" of the map centered on the Caspian Sea in the former Soviet Union. Once you have the map centered on the Caspian use the *magnifying glass tool* at the bottom left to magnify the same area (i.e., move the triangular pointer in the lower left of the screen from its "home" position in which the apex of the triangle is midway between the last set of tick marks on the scale toward the icon of the magnifying glass with the plus sign on it). Stop at the midpoint of the adjacent set of tick marks from the home position.

1. What is the scale now?

We can note that the legend bar (about nine centimeters in length) now represents 3,000 km rather than the 6,000 in the previous representation.

2. Does that make this map a larger scale than its predecessor?

3. What is the value shown when you click on the mouse (left side) when the scale on the legend goes to 3,000 km?

4. What is the title of the map shown on the screen at the scale with the 3,000 km legend bar?

Every time we move the sliding point to another adjacent tick mark on the lower left-hand scale on the magnifying glass we are honing in on an increasingly smaller area on the ground and thus we are viewing a progressively larger scale map.

5. How does the title change when you move the sliding triangular-shaped pointer to midway between the next set of tick marks (i.e., an increase in the scale such that only 1,500 km appears on the legend bar)?

6. What happens when the scale is approximately doubled again by moving the sliding triangular indicated midway between the next set of tick marks (i.e., when the legend bar equals 800 kms)? Does the title of the map change again?

7. What happens to the map title when the scale is enlarged once again? At this point the 9-cm legend bar represents 400 kms.

By now you make have realized that you can point and click rather than sliding the pointer and accomplish the same tasks much more quickly.

8. What country or city is shown when the scale on the legend bar (bottom right side) is moved to 150 km?

9. 80 km.?

10. 40 km.?

Now let's experiment with another tool contained on Encarta. Restore the map to its world scale again (with at least 3000 kilometers shown on the legend bar). Using the *Hand* symbol, rotate the small globe in the upper right by dragging your mouse as if you were actually running your fingers over a globe in the classroom. Rotate until you can clearly view both sides of the Pacific Rim simultaneously (i.e., such that Japan and China are on the left-hand side and California on the right). Now, suppose you wanted to fly the most direct route from San Francisco, CA to Seoul, Korea. Make sure you know approximately where your origin and destination points are located before proceeding. Now, pull down the menu at the top of the screen called *Tools*. Within the *Tools* utility, click on the *Measure Tool*. Locate San Francisco and click on it as the origin. Now find Seoul and click on it as the destination.

11. How far is it (in kilometers) from San Francisco to Seoul?

12. How would you describe the shape of the line that the Measure Tool has drawn?

13. What island group would you fly over to get from San Francisco to Seoul via the shortest path? Does this surprise you? Why or why not?

14. When a pilot talks about a Great Circle Route do you think you understand what that means now?

Keywords: **map scale, Mercator projection, Great Circle Routes**

Chapter 2: Cactus, Cowboys, and Coyotes: The Southwest Culture Region

Postcards from Encarta: Wish You Were Here!

Part III of this chapter focuses on "Regional Imagery," and asks you to collect postcards that are representative of the region in which you live. When you tackled that assignment, you may have been surprised by how others view the place where you live. If you read the fine print on the back of most postcards, you probably noted that often the photographers, artists, and publishers are not from your hometown and more often than not may not even be located in your country. Were the postcards from your town representative of your town or region? Did you learn anything new from the captions on the back of the postcards or did you find any editorial errors? Perhaps there are landmarks you would have liked to have seen memorialized on a local postcard? This activity will give you a chance to produce your own postcards that you believe best represent the geography of your town.

PART I: Greetings From North America!

Congratulations! You have just been named the CEO of a postcard company and for your first pet project you decide to establish an entire line of geographic postcards based on Joel Garreau's "Nine Nations of North America." Using Figure 2.2, "Nine Nations of North America, A Newspaper Journalist's Perceptual Regions," from your text as a guide, you will create a series of original postcards representing each of Garreau's nine vernacular regions. Follow these steps for constructing your personalized postcard collection.

1. Use the *Find* tool and create a map of *North America*. Using Figure 2.2 in your text as your guide, zoom in on the locations of each of Garreau's regions by clicking on what would be the upper left (northwest) corner of the region's boundary and drawing a box on the map. Click to zoom in to create a new map that best represents the region you wish to examine. The new map won't have the exact boundaries as Figure 2.2 but you should try to include as much of the region as possible. An alternative to this approach is to go directly to the maps of states and or provinces that are included in each region. If you choose this method just be sure the part of the state you are looking at is included in the corresponding "nation."

Joel Garreau's Nine Nations of North America

Ecotopia	**The Empty Quarter**	**MexAmerica**
The Bread Basket	**Quebec**	**The Foundry**
Dixie	**New England**	**The Islands**

2. Once you have a map that closely resembles one of the "Nine Nations," go to *Sights and Sounds* and find an image that you believe most closely represents what you would like to see on your postcard. You should look at the slide show for each of the states and or provinces in the region before you make your decision. One you have

selected an image you believe best illustrates each area, print that slide, trim the edges and narratives and set it aside for the next step--captioning.

3. When you come to the area in which you live, *select two images*---one that you think best represents what Garreau was trying to say about the region, and one that you would like to see on a postcard rack in a local store (select one from your state or province).

4. For each slide, you will need to come up with a short caption that in just a sentence or two sums up what the viewer needs to best understand the postcard image. Do not use the same information from *Encarta Virtual Globe '99*--be creative! Cut and paste your caption on the upper left hand corner of the postcard, where you could normally find a brief description.

PART II: Return to Sender

You should now have a collection of ten postcards---one for each of Garreau's "Nine Nations of North America," and one extra from your home state or province. If you were mailing these to family or friends, on which of these postcards would you write "wish you were here" and on which would you say "be glad you are someplace else" and why?

Keywords: **culture region, Joel Garreau, Nine Nations of North America, postcards, vernacular region, perceptual region**

Chapter 3: Tracking the AIDS Epidemic: Diffusion Through Time and Space

Digging to Chinatown: Relocation Diffusion in Action

Have you ever wondered how ethnic neighborhoods are established or why some sections of cities are dominated by a particular culture group while others are defined in other ways? A look into urban ethnic neighborhoods will give you some insights as to the end result or relocation diffusion, chain migration, and prejudice.

When someone mentions "Chinatown," what location springs to your mind? If you first thought of San Francisco it may be because you are from California or the western United States or perhaps you have seen this neighborhood depicted in the media. If you live in or near British Columbia, your idea of Chinatown might be Vancouver, the second-largest Chinese community in North America---after San Francisco. Or perhaps you call Australia home and the city of Brisbane comes to mind for its large Chinese community. Let's take a look at all three "Chinatowns" and explore how diffusion dynamics have resulted in the creation of these distinct ethnic neighborhoods. For additional background on how diffusion has affected global demographics, call up *Encarta's Virtual Globe '99* and take a few minutes and read "*Human Migration*" in *Global Themes*. You can access this via the *Features* option and then by clicking on "*The World of People*," and finally "*Human Migration*."

PART I: The Streets of San Francisco

Our first stop on our Chinatown tour is San Francisco, California. Create a map of the city by typing "*San Francisco*" in the *Find* box. Once you have a map of the city, click on *Geography* and read a short summary of San Francisco's history. You can make a street map of San Francisco that will show you the exact location of Chinatown by entering "*Chinatown*" in the *Find* box and selecting the second California entry (the first is for Los Angeles, which also has a large Asian community). To view an image of a Chinatown street scene, select the *Sights and Sounds* option.

1. What is the sequence of immigration to San Francisco beginning in 1776 (i.e., which ethnics groups settled there since the 18th century)?

2. What events spurred great numbers of immigrants to the San Francisco area during the 1850s? Why do you think these circumstances drew large numbers of Chinese and other immigrants to the city?

PART II: Vancouver

Our second stop on our Chinatown tour is Vancouver, British Columbia. Go to a map of the city using the *Find* command and read a short history of Vancouver by clicking on

Geography. Next, view the slide, *"Vancouver's Thriving Chinatown"* by selecting *Sights and Sounds*.

3. What attracted Chinese immigrants to Vancouver during the 19th century?

4. Are there any parallels between the settlement of Vancouver's Chinatown and that of San Francisco? Explain.

PART III: Chinatown Down Under

The final stop on our global Chinatown tour is Brisbane, Australia. Enter *"Brisbane"* in the *Find* box and bring up a map of the city. Read the brief description of the area by selecting *Geography* and then view a slide of Chinatown by clicking on *Sights and Sounds*.

5. What is different or similar about the immigration history of this Chinese neighborhood to those you have studied in San Francisco and Vancouver?

6. What locational similarities and differences does Brisbane's Chinatown share with the other two neighborhood examples in this activity? Why do you think this pattern emerged over time?

PART IV: Fine Chinatowns

You have been a virtual visitor to three ethnic Chinese neighborhoods in the world but there are many more to explore. Just out of curiosity, type in *"Chinatown"* in the *Find* box and, under the headings below, list other cities that have their own Chinatowns.

UNITED STATES   CANADA   OTHER COUNTRIES

7. As you review the list you created, briefly discuss in regional terms why you think this pattern of Chinese neighborhoods exists. Are there locational similarities? Does the size of the city have an influence or perhaps are some historical connections to the diffusion pattern?

8. Why do you think Chinatowns have survived for over 100 years? Do you believe these neighborhoods today serve as foci for Asian immigration and culture? For example, would you expect a Chinese immigrant to naturally gravitate to a Chinatown-type neighborhood today as previous migrants did in the 19th century. Why or why not?

Keywords: **diffusion, immigration, Chinatown, San Francisco, Vancouver, Brisbane**

Chapter 4: Newton's First Law of Migration: The Gravity Model

A Matter of Some Gravity: World Population Cores and Distance Decay

The book does a wonderful job of illustrating the gravity formulation based on Newton's law of the attraction of heavenly bodies. In the Newtonian formulation, the exponent to which distance is raised is two (i.e., the relationships are distance squared ones). In more pragmatic applications, variants of the classic gravity formulation have been developed to forecast forms of human interaction. This interaction can include, but is not limited to, the migration of people, the movement of commodities and even the selection of marriage partners. Instead of distance being squared in the denominator of the classical gravity formulation, the exponent to which distance is raised is sometimes treated as a parameter to be fitted—the lower the exponent value the less the frictional effect of distance on that form of interaction and vice versa.

What can Encarta do to illustrate the gravity model given that large scale subnational data are not readily available (although they might be tapped from appropriate *Web Links* as shown in other activities throughout this *Activity Guide*)? How about simply confirming whether a gradation of population from key core areas at the world scale is really visible? If so, such confirmation should lend credence to the extraordinary power of the deceptively simple gravity formulation. Clicking on *Features*, pull down the *Map Gallery* tool and go to *Earth at Night.* This dramatic "map" is ready a composite of many satellite images taken on various clear evenings so that cloud cover is at a minimum throughout the globe and it is night time everywhere simultaneously. The most obvious features on the map are the lights of the urban conurbations from giant behemoths like New York or Tokyo or Mexico City to smaller cities and towns dotting the landscape. The distribution of population void areas (the dark patches on the satellite image) speaks volumes about where people are not just as the light patches tell us where they are.

1. Does the pattern of human settlement throughout the world in any way illustrate the gravity model?

We think you'll find that it does although it helps to know a little about the settlement history, the resource base, and the barriers to human settlement to really understand the distribution of world population. Dragging your mouse to move the globe (i.e., the small globe in the upper right hand corner of the screen), center the globe on Europe and then use the *Magnifying Glass Pinpointer* to increase the geographic scale (the legend bar should show 2,000 km. or 1,500 km.)

2. Would you say that Europe has a core area?

3. Where is it exactly?

The Dutch call this area the Randstat and it consists of Europe's largest deep water port of Rotterdam in addition to Amsterdam and several other large cities which have

69

coalesced together into a major conurbation. As you can see, the core area seems to extend across the Channel to include the United Kingdom as well.

4. Where within the UK is the core area centered?

The core area extends into the former West Germany and a heavy industrial area known as the Rhine-Ruhr Basin. Can you find that portion of the European core? Beyond the core, the population density seems to decrease with increasing distance from it just as the gravity model would predict. There are, however, some significant outliers—large places or conurbations that are not contiguous with the core per se. The Paris Basin in north central France seems to be one such outlier.

5. Can you name some others within Europe?

6. Why do you think that population is concentrated at these nodes?

Now, spin the small globe (by dragging the mouse) and focus on Asia.

7. Why do you think that Japan with its approximately 125 million people stands out so brightly whereas China with almost ten times its population appears only modestly lit up?

8. When we move away from the coastal areas and the river valleys of China, the bright lights become almost nonexistent. Why is that?

Finally, spin the globe so that it is centered on North Africa, specifically on Egypt.

9. What is unusual about the pattern of population in Egypt?

10. What factors account for this concentration and unusual distribution?

Cairo is Africa's largest city with over 10 million people in the greater metropolitan area. Can you spot it easily on the nighttime satellite image? Several years ago, the American geographer Waldo Tobler, building on the previous work of Swedish geography Stig Nordbeck, thought that satellite imagery might be used to conduct a type of census and he used Egypt to illustrate his method. Like the gravity model, Tobler based his model on a natural law too. In his case, the law pertained mainly to living organisms and comes from the field of biology. It is called the law of allometric growth. Simply stated, the law argues that the growth of an organism's limbs and organs is proportional to the overall growth of the organism. Hand and feet don't just suddenly start growing faster than the rest of the body. The entire body grows in a proportional manner. Using this law, Tobler was able to relate the size of the built-up urban area as a surrogate for the actual population living there. Of course, a distance decay effect was assumed with more people living in the central area of the cities and the density of populations tapering off near the periphery. Applying this natural law to the distribution of Egypt's population, Tobler estimated the population of the country within the acceptable degree of error of the official census that cost millions of dollars to conduct. Tobler's method cost mere hundreds at the most—a real boon to financially strapped third world countries who might find it difficult and expensive to conduct a thorough census.

11. Do you think a census conducted using Tobler's method would be as accurate in the rainforests of Sumatra as it was in the desert of Egypt? Why or why not?

12. Is it comforting to you to know that human behavior (in the aggregate anyway) is so predictable that natural laws such as Newton's law of gravitation or the biological law of allometric growth do a remarkably good job of predicting our behavior?

13. Does it make you want to rebel against the norm just so you will not be considered so predictable?

14. Would your capricious, aberrant behavior really do much to change the fundamental nature of these natural laws or their sociological and geographical derivatives?

15. Are you a residual (i.e., outlier) in the grand scheme of things or a well-explained case?

16. If geographers can predict your spatial behavior with a remarkable degree of accuracy, does that prediction in any way deny your free will to act as you want to?

Keywords: gravity model, law of allometric growth, Egypt, Waldo Tobler, satellite imagery, core area, urban agglomerations (conurbations)

Chapter 5: Trapped in Space: Space-Time Prisms and Individual Activity Space

My Prism Can Be My Prison

The notion of time-geography, the very choreography of our existence, is a fairly new one first discussed by Törsten Hägerstrand and his Swedish associates in the 1970s. It was that group of geographers who saw great potential in charting the daily life path of the average citizen. Once you get beyond the technical literature involving space-time prisms, space-time paths, coupling constraints, dioramas and the like, the concept is a fairly simple one—you can't be in two different places at the same time. And, furthermore there may be opportunity costs foregone because of our relative immobility. When daily logs were kept we learned that working class people may be more constrained by their time-space prism (prison?) than their upper middle class counterparts because they lack access to high speed mobility systems. We have also learned that there are gender and age-related differences that can make a person relatively confined to a limited geographic space. For example, a young wife with several pre-school children might be precluded by economic circumstance or personal preference from participating in many of the activities going on outside the home.

These concepts, while important, are individualistic and difficult if not impossible to illustrate using *Encarta's Virtual Globe '99*. But, the related concept of the time-space convergence, first introduced into the geographic literature by Donald Janelle, can certainly be illustrated using *Encarta*.

In an early historical atlas of the United States, Wright and Paullin developed an interesting series of maps to illustrate the speed with which a message could move into the interior of the country at three different time periods. The earliest time period chosen was 1800. At this time, overland travel was very expensive, arduous and slow. On the other hand, sailing ships plied the waters between port cities in the United States bringing cotton from Charleston, SC to Boston in exchange for shoes or woolen goods. Thus cities along the eastern seaboard were interconnected. In 1800, it would have taken a full six weeks for a message or a shipment of commodities to reach what was to become the city of Chicago at the southern tip of Lake Michigan. At the time it was Ft. Dearborn--a military outpost designed to protect fur traders and other hardy souls in that region against attacks by hostile Indian tribes in the area. By 1830, the steam engine had been added to packet boats that plied the navigable inland waterways opening up vast hinterland trade areas for expansion. Especially important was the opening of the Erie Canal in 1825. This canal opened up a sea level route from the port of New York into the Great Lakes system via the Hudson and Mohawk Rivers that formed a corridor throughout upstate New York. The steam engine led to the growth of cities along the corridor such as Rochester, Utica, Troy and Schenectady. Despite the advances in technology, it still took three weeks for a letter or a shipment of commodities to reach Chicago. The city of Chicago was incorporated three years later and by the 1880s was the second largest city in the United States. In 1857, the last year examined by Wright and Paullin, the first railroad bridge across the Mississippi had been constructed, the first transcontinental railroads with a standardized gauge of track was less than a decade away and a message from New York could now reach Chicago in only two days. The interesting isochronal maps (i.e., showing lines of equal travel time away from New York

72

City) of the historical atlas illustrate well the time-space convergence. From six weeks in 1800 to three weeks in 1830 to two days in 1857 a message could be sent from New York to Chicago .

Now click on *Find* in Encarta and type in *"New York."* Using the mouse to drag the map around a bit, making sure that New York City is approximately in the middle of a map in which the legend bar goes to 80 km. This is what a driver for overland transportation system would have had to face in 1800.

1. What are some of the barriers to movement overland that a driver would have to face?

2. What physical features are located west and northwest of the city? The areal extent of the map is about as far as one could go in two or three days in a horse-drawn wagon.

Now, using the *Magnifying Tool*, click on the *minus icon* such that the map is still centered on New York City but the map legend in the lower right corner of the screen now extends 150 km. This is just about what happened between 1800 and 1830 according to Wright and Paullin. The time it took to go from point A to B was halved. Thus, a steam packet or improved overland vehicle such as the iron horse, could travel to the places shown on this map in two-three days.

3. To what states does the map at this scale reach?

The time-space convergence sped up between 1830 and 1857 much more rapidly than it did in the 1800-1830 era. Now click the *magnifying glass* (negative sign) twice such that the legend that is showing extends to 800 km. Chicago is now at the periphery of the region shown on the map taking two to three day delivery time as it did in 1857. What are the most far-flung states from New York that can still be reached in two days? In less than sixty years the convergence went from 80 km in a fixed time period (2-3 days) to 800 km, a tenfold speeding up (convergence). Did your neck snap in the process due to the G forces?

4. Can you think of improvements in transportation and communication technology that have contributed to the space-time convergence that we have experienced in our affluent and developed country?

Keywords: space-time prisms, time-space convergence, time-geography, Törsten Hägerstrand, Donald Janelle

73

Chapter 6: Help Wanted: The Changing Geography of Jobs
Around the World from Pre- to Post-Industrial

Figure 6.1 in the text has been used to illustrate how a particular country such as the United States had made the transition from an agrarian society to a post-industrial one. In the early years of the United States' history, the economy was heavily dependent on the primary sector of the economy--the jobs that are directly reliant upon raw materials from the earth or sea. Later, manufacturing (i.e., the secondary sector of the economy involving the change of those raw materials into more useful products thus adding value to them in the process) became the backbone of American emergence as a world economic power. Now, the United States is in a post-industrial phase in which the service industries (both the more routine tertiary sector and the higher-order and more highly professional quaternary sector) are dominating the jobs Americans perform. So, Figure 6.1 can be interpreted in a longitudinal sense for a particular country's economic transition. But, can the diagram also be used in a cross-sectional sense? That is, for a particular time period would we be able to find different countries in the world at different stages in this transition? If so, could the stage a country finds itself in be used as an indicator of level of economic development?

Let's use *Encarta's Virtual Globe '99* to find out. We'll focus on five countries that are at different stages of development—*Bangladesh*, a desperately poor country in South Asia that some development experts have even labeled a "fourth world basket case" because of its seemingly intractable problems. Then, let's focus on *Thailand*, a third world country in Southeast Asia that is currently emerging from its traditional roots to become a very fast-paced economy. Third, let's examine *Brazil*, a huge country with a great, albeit spotty, natural resource base and development potential; a country that development experts refer to as a newly industrializing country (NIC). Of course, we shouldn't overlook the *United States* and for comparison let's look at *Switzerland*, a country in Europe with a standard of living (as measured by per capita income) even higher than our own. Is their employment base that different from our own? Can we learn some lessons from Switzerland's obvious success?

Using the *Find* option in Encarta, type in the names of each of these five countries and then click on the sidebar entitled *Facts and Figures*. To move from one country to the other simply go to the pull-down menus at the top of the screen and pull-down on *Find* and then click on *Places* and simply type in the name of the next country you are interested in. Create a table by scrolling down the *Facts and Figures* section until you come to *Economy* and record the following information for each of the five countries—per capita GDP (gross domestic product), the proportion of the labor force engaged in agriculture (primary sector), industry (secondary sector), and services (tertiary and quaternary sectors).

1. What appears to be the relationship between per capita income (GDP) and the percentage of the labor force engaged in agriculture?

2. What is the most unusual aspect of Switzerland's economic structure?

3. Even though Switzerland's per capita GDP is half as much as that of the United States, it's economic structure is a bit unusual. What is the unusual aspect?

Only countries in Africa such as Angola and Botswana have a higher percentage of their labor force engaged in manufacturing than Switzerland. What kind of manufacturing do the Swiss engage in? Let's take a closer look with *Virtual Globe '99*. Using the *Find* command, type in '*Switzerland*' and return to the *Facts and Figures* sidebar for a moment.

4. What are Switzerland's three leading exports?

Now click on the sidebar called *Sights and Sounds* and click on the slide entitled "*Making watches in Switzerland*".

5. Where within Switzerland are most of the fine time pieces made?

Sometimes a time honored tradition of craftsmanship can negatively impact market share. Before 1960, some 90 percent of the watches in the world used Swiss movements. It was, in fact, the Swiss who invented and developed the prototype for a quartz watch. But that technology did not fit into the methods then used to produce watches in Switzerland and the right to fabricate and commercialize the quartz watch was sold to the Japanese government. Because of the close alliance of business and government in "Japan, Inc.", soon Japanese watchmakers such as Seiko, Citizen, and Pulsar were dominating the world market for watches. Japanese watches were reliable and inexpensive. Swiss watches are wonderfully crafted, reliable and expensive. The Swiss fought back with the Swatch watch, a quartz watch produced in many designer colors and patterns but the Swiss have never regained the market share they once enjoyed.

Now click on the sidebar called *Society* and then click on the section called *Economy*.

6. What country is Switzerland's most important trading partner?

7. Can you think of another country that might fit the opening statement in the Economy section: "Despite a lack of natural resources, Switzerland has a strong economy?"

Reflect for a moment on the difference between the economies of the *United States* and *Switzerland*. The United States used to have a stronger industrial base than it does now. Many of the former domestic manufacturing jobs have been outsourced to third world

75

countries where the products can be produced at much lower costs than in the United States largely because of the labor cost differential between the United States and the third world. Why aren't Swiss manufactured products outsourced to the same degree? Switzerland, like Germany and many other European countries, has a very strong program of craftsperson apprenticeships. Should such programs be implemented in the United States too so that we keep our technological and skill-based edge in manufacturing? Why or why not?

Keywords: economic sector, gross domestic product, Bangladesh, Thailand, Brazil, United States, Switzerland, watches, manufacturing

Kuby, et al., *Human Geography in Action*

Chapter 7: The Hidden Momentum of Population Growth
Another Type of Egyptian Pyramid

You have undoubtedly seen pictures of the colossal pyramids at Giza in Egypt, one of the seven wonders of the ancient world, and still a major tourist destination despite recent terrorist attacks. But there is another pyramid that looms large in the minds of many Egyptian planners and government officials—the demographic pyramid (i.e. age-sex cohort diagrams). These diagrams are usually referred to as pyramids even though their shape can vary dramatically from that of a pyramid. Just look at the "pyramid" for Spain, a country that now possesses the lowest birth rate of any European country (incredible for a country that is nominally Roman Catholic but true nonetheless) as pictured in Figure 7. 4. Only those age-sex cohort diagrams for the Philippines and Zaire (now the Democratic Republic of the Congo) look very pyramid-like with a broad base and then tapering to an apex.

What about Egypt? Wouldn't it be ironic if its age-sex cohort resembled the Great Pyramids at Giza?

1. If it did, what would be the long-term consequences of such a shape?

Let's turn to *Encarta* to learn more about the fascinating land of Egypt and some of the ways it has come to terms with its extremely high physiological density (i.e., the number of people per unit of arable land). Since most of the land area of Egypt is desert, the population of over 62 million is forced to live along the Nile River floodplain, its tributaries, a few oases, and the delta it forms with the Mediterranean. The Nile is truly Egypt's life blood and has been over the millenia.

Using the *Find* command in Encarta type in *'Egypt'*. First let's examine some demographic data pertaining to Egypt by clicking on the sidebar menu called *Map Styles* and then clicking on *Statistical*. At the bottom center of the screen is a box entitled *Choose Statistic*. Click on it and bring up the menu of choices and within the many categories of data sets contained therein, click on the one marked *Population*. If you've done this correctly, choropleth maps by country will display the variable you've chosen. By simply placing the mouse on the country of interest, the value of that variable for Egypt will be revealed.

2. What is Egypt's birth rate (1997)?

3. What is Egypt's death rate (1997)?

4. What stage of the demographic transition does Egypt most likely fall into?

Click on and record the value of the following population variables for Egypt—*fertility rate (1997); population growth rate (1997); population growth rate, urban (1995-2000); population aged infant to 4 as a share of total population (1997); population aged 5 to 14 as a share of total population (1997)*; and *population aged 65 and older as a share of total population (1997)*. Notice where within the data array, the values for Egypt fall (of the countries reporting data). That's right. Egypt is just about smack dab in the middle. And, for a country with a large population base to begin with, that is a good place to be. We're sure that many Egyptian planners and government officials wish that rate of natural increase were even lower than it is now. But, Egypt has lower rates of increase and lower fertility rates than many of its neighbors in North Africa and Southwest Asia. Still, 3.5 children per woman of childbearing years in 1997 is quite high by Western standards. It is, however, considerably lower than the figure of five children born per woman of childbearing years in Egypt in 1990.

What about the base and the apex of the population pyramid? The rate of juvenility is defined as the proportion of the population that is considered too young to be on their own (i.e., the dependent population less than 15 years of age).

5. What proportion of Egypt's population would be considered dependent? (HINT: Add the population aged infant to age 4 as a share of total population to the population aged 5 to 14 as a share of the total population)

6. How does this figure for Egypt compare with that for the United States?

Now let's examine the apex of the pyramid.

7. What proportion of Egypt's population would be considered elderly (i.e., aged 65 and over)?

8. How does this figure for Egypt compare with that for the United States?

But, keeping with the cliché that a picture is worth a thousand words, let's look at Egypt more closely. Click on the sidebar marked *Sights and Sounds* and turn on the speakers if your computer is equipped with a sound card because you will be treated to at least four different types of music along the way. Focus first on the slide entitled '*Egyptian Farmland*'.

9. If it weren't for a government edict mandating that some food crops be grown, what crop would probably increase the most in acreage under production?

10. Since land was recently expropriated from wealthy landholders and redistributed to the peasant farmers, what is the average sized holding in Egypt?

Now turning to the slide called "*Nile River at Aswan*", please answer the following questions.

11. When was the Aswan High dam completed?

12. Why was it built?

13. Did its construction allow Egypt to add to its arable land resource base?

14. What antiquities had to be moved as a result of the dam's construction and the subsequent submergence of previously dry land?

15. According to the slide entitled "*Farming in the Nile River Delta*" what percentage of Egypt's population farms for their livelihood?

16. What factors account for Egypt's high agricultural productivity?

17. The caption to "*Windswept Desert*" says what Egypt's annual naturally-occuring rainfall is. What is it?

18. Focusing on the "*Aswan High Dam*", what large body of water was created south of the dam?

19. Who or what is Nassar?

20. What have been the benefits (if any) of building the high dam on the Nile?

21. What are the negative consequences (if any)?

22. How old is the "*Great Pyramid at Giza*"?

23. How many stone blocks were used to build it?

24. At the "*Al Fayyum Oasis*", we are introduced to a different type of extractive activity— aquaculture. How does the aquaculture practiced at Birkat Qarun differ from catfish farming in Mississippi or tilapia ponds in southeastern Asia?

25. How is the salinity level kept relatively constant?

26. We can't let you miss the "*Giza Sphinx*". From what material was it created?

27. How long ago?

28. In the slide simply called "*Village Life along the Nile*" it is stated that "Rural life is dependent upon the contributions of family members." If there is no national system of social security or old age pensions, would this be a reason why the fertility rate is still relatively high?

29. In the slide "*Expansive Lake Nassar*" we see once again the potential environmental consequence of building a 300 km. lake in a desert environment. Why do you think that simple evaporation is such a concern?

30. Finally, take a sweeping view of the largest city in Africa—Cairo in "*Metropolis of Cairo*". What is you general impression of the city based on that photograph? Is Cairo the future for a country that is now 80 percent rural?

Keywords: **fertility rate, population pyramids (age-sex cohorts), Egypt, Aswan High Dam, Lake Nassar, Nile River**

Chapter 8: From Rags to Riches: The Dimensions of Development
Ecotourism: It Isn't Easy Being Green--But It's Profitable

Picture this. You have worked hard all year long at school and are more than ready for spring break. Where do you go? Cancún? Disney World? Myrtle Beach? No! You opt for a week of trekking through tropical vegetation, coated with insect repellent, somewhere in Latin America so you can single-handedly save the rainforest. Huh? That's right---you are an ecotourist and are more than willing to spend your hard-earned dollars (and next semester's tuition) to experience some of the world's most fragile and endangered environments. Why would you rather commune with nature than take a ride on Space Mountain? Well, maybe you have a fear of heights or just maybe you are one of a growing number of people from the more developed countries (MDCs) joining a global movement toward ecotourism. What is ecotourism and what is its role in economic development? Hopefully we will answer those questions in this activity and perhaps spur your interest in widening your vacation horizons.

To get you started, enter "ecotourism" in the *Find* box. When the *Contents* window loads, click on "ecotourism, glossary," and read the definition that is listed. You may wish to print this off for future reference. While you have the *Contents* box handy, take a look at the locations listed that have something to do with ecotourism. To make your work a little easier for these activities, you might want to click on the "keep find open" box just below *Contents* (Hint: as you move from screen to screen the *Find* and *Contents* boxes will disappear. To bring them back just click on *Find* at the upper left corner of your screen). In the following, we will travel to three of the places listed. You can use this list to quickly go to the locations where we will be doing some virtual ecotourism.

PART I: Jamaica To the Islands Yet?
Jamaica has everything from coastal plains to rugged mountains, but as of yet, has not been a haven for ecotourism. To find out why Jamaica has not taken advantage of this lucrative sector of the tourism market, click on "*Jamaica, Land and Climate*" in the *Contents* box. Read the descriptions of location, topography, climate and environmental issues then consider these questions.

1. What role has the Jamaican government played in protecting the natural environment? How has this affected ecotourism?

2. What are some of the major threats to the Jamaica's island ecology? How might these factors impact tourism of all kinds?

3. Why does Jamaica have such rich biodiversity and "endemism?"

4. If you were hired by the Jamaican tourism board, what solutions would you offer to solve some of the environmental problems and to develop ecotourism?

PART II: Panama: Hats Off to Ecotourism

Panama, much like Jamaica, has great potential for ecotourism but has yet to capitalize on this growing segment of the travel market. Compare what you have learned about Jamaica to Panama by selecting "*Panama, Land and Climate*" in the *Contents* box and reading about the country's physical geography and environmental issues.

5. How does Panama's site and situation account for its vast biodiversity?

6. Why are Panama's coastal reefs, mangrove moist forests, and ecofloristic zone are endangered?

7. Why do you think the Panamanian government has been more successful in protecting fragile environments than that of Jamaica?

8. Which country has the greatest potential for developing ecotourism---Jamaica or Panama and why?

PART III: Papua New Guinea: Seeing the Tropical Forest for the Trees

Our third stop on our tour is Papua New Guinea (PNG). Begin your tour by reading "*Papua New Guinea, Land and Climate*" which you can select from the *Contents* box.

9. What are the two biggest threats to Papua New Guinea's vast biodiversity?

10. While you are looking at "*Environmental Issues*," in the *Land and Climate* section, click on *Statistics* "*Forest, annual rate of deforestation*." Look at the rates of deforestation for PNG, Jamaica, and Panama. Which is losing the greatest percentage of forest cover and which one is losing the least? How do you explain the numbers for PNG and what are the prospects for the future?

11. Why do you think the PNG government is promoting ecotourism as economic development?

PART IV: Before You Pack Your Bags...

Now that you have had a chance to "visit" several places where ecotourism is developing you need to decide where you would want to go if given an opportunity to visit Jamaica, Panama, or PNG. To check out the details of traveling to these places visit these web sites for each place by clicking on *Web Links* at each country's basic page on *Encarta*.

- *Destination Jamaica, Destination Panama*, and *Destination Papua New Guinea*
- *U.S. State Department Travel Advisory* (for each of the three countries)

12. After reviewing the information for each and based on what you have learned about environmental conditions in each country, which of the three places would you prefer to visit and which would be your least favorite destination as an ecotourist? Why?

Keywords: **ecotourism, economic development, biodiveristy, environment, Jamaica, Panama, Papua New Guinea**

Chapter 9: Take Me Out to the Ball Game: Market Areas and the Urban Hierarchy

Towns in Iowa: Central Places and a Whole Lot More!

As you might be able to gather from some of the studies cited in Chapter 9, Iowa is considered the classic central place testing ground. It is relatively flat to gently rolling. While not the isotropic plain of the classical theory it comes pretty darn close except for the slightly more rugged northeastern part of the state that is part of the Driftless Area, a region missed by the most recent glaciation. What does the glaciated landscape of Iowa look like? You can find out by typing '*Iowa*' in the *Find* box. Then go to the sidebar marked *Sights and Sounds* and you will see two slides of the Iowa landscape. The first is a typical farming scene ("*Iowa's Riches*"). Note how flat the land is in all directions. What crop do you think the farmer is driving his tractor through? If you said soybeans, you would be right. Soybeans are now second only to corn in importance to the state. Soybean acreage has increased steadily since the 1940s in Iowa eclipsing hay and oats to become the second leading agricultural crop. Why? Because soybeans have a relatively short growing season, as a legume they produce their own nitrogen fertilizer fixed by bacteria that attach themselves to the roots of the plant. They have dozens of uses from meat extender to cattle feed. They have never been subject to an acreage allotment. And, there is a huge foreign market for soybeans and soy products in the Pacific Rim. The second slide shows an area of Iowa close to the floodplain of the Mississippi River in the eastern part of the state ("*Mississippi Farmland*") where the effect of glaciation during the most recent Ice Age is quite apparent.

Certainly glaciation is one reason for the look of the land in the state of Iowa but there is another factor that has also had a profound effect on the settlement pattern of the state—the township and range survey system. This survey system was used to divide the state into political units called townships and counties as part of the Northwest Ordinance Survey of 1785. This was long before Iowa was granted statehood (1849) and long before very many people lived there. Iowa was simply part of the Northwest Territories, territories north of the River Ohio. How can something as seemingly mundane as how land parcels are platted on maps even before any significant population lives there affect the distribution of central places? The typical county was square in shape consisting of 16 townships. A township, in turn, consisted of 36 one square mile parcels of land called sections. A section of land (i.e., one square mile) consists of 640 acres. During the Homestead Act of 1862, parcels of land not yet settled (mostly in the northwestern part of the state—the last to be settled) were allocated to people who wished to farm the land in quarter sections (160 acre parcels). So, the average farm size was about 160 acres (one-quarter section). All of the farm-to-market roads were oriented along the township and range boundaries. You didn't want to cut across a farmer's acreage if at all possible. Every township declared one centrally located section as set aside for the one-room school and other township functions. Likewise, every county seat town was centrally located within the county boundaries. Since there are 99 counties in the state, one might assume that the 99 largest cities would be the regularly spaced county seat towns. And, for the most part you would be correct. There is, of course, some degree of suburbanization around many of the larger cities which leads to a bit of clustering (i.e. urban agglomeration) but still the most remarkable feature of the Iowa landscape would be the regularity of spacing of the major central places and, by implication, all levels of the central place hierarchy from the smallest hamlet to the state capital of Des Moines. But, you need to prove this regularity for yourself. Using the map of Iowa with the scale shown on the bottom right going to 150 kilometers, let's measure the distance between seven largest central places in the state and their nearest neighbor of the same order. Those seven are Des Moines, Cedar Rapids, Waterloo, Davenport (not labeled per se, but part of the Rock Island and Moline, Illinois Quad Cities area on the border of eastern Iowa and western Illinois), Sioux City, Council Bluffs (right across the river from the much larger Omaha, Nebraska) and Dubuque. For example, Cedar Rapids' nearest neighbor of comparable hierarchical rank is Waterloo. What is the distance that separates them? Use the *Tools* pull-down menu and click on *Measure Tool*. Place one end of the Measure Tool on Cedar Rapids and the other on Waterloo and record the straight-line distance (measured in kilometers that separates them). Your answer should be about 84.6 km. Click the cross-hairs on the origin and again on the destination to take a reading. Now record the values for the other six pairs of cities.

1. What was the arithmetic average of your result? HINT: Add all seven nearest neighbor distances and divide by the number of city pairs considered, in this case seven.

If your average was around 120 kilometers, you did the exercise correctly.

2. Why do you think that on average the distance between major centers in the eastern part of the state is less than that in the western part?

In southern Germany, Walter Christaller, the geographer responsible for first developing central place theory, found that the average spacing of his largest urban centers (excluding the regional Bavarian capital of Munich) was about 36 kilometers.

3. How do you account for the differences between the results obtained by Christaller and your own?

4. What kinds of central place goods and services might you expect to find only at the highest level of the central place hierarchy in Iowa?

5. Brian Berry and his associates used hospitals to represent the town level function in their study of southwestern Iowa (Figure 9.1 in your textbook). What would you use to represent the city level?

Keywords: Iowa, Northwest Ordinance Survey, township and range survey system, Homestead Act, central place theory, glaciation, Driftless Area

84

Chapter 10: Reading the Urban Landscape Through Census Data and Field Observation
All Cities Great and Small

Since the late 1800s suburbs have drawn populations, investment, retailing, industry and service employment away from cities. Today, some of these suburbs are virtually cities in their own rights, cities where the downtown is the shopping mall, and where the population growth rates outstrip that of the city that spawned the suburb in the first place. But not all urbanized areas have experienced rapid growth in the past few decades. In fact, some cities that were considered to be the leading edge of urbanization during the early industrial era are struggling to maintain their populations today. In this activity you will have a chance explore world urbanization as well as track the growth and decline of American cities and perhaps your own hometown.

PART I: Urban Globetrotting
Before we begin our data analysis, we need some background information on suburbs and the history of world cities. To access this material type "*urbanization*" in the *Find* box. When the *Contents* box fills up, select the first entry, "*Urbanization and Cities, World Themes.*" Read this section and consider the following questions.

1. How long ago were the first cities established and what has brought about rapid urbanization in past the 200 years?

2. How do the United States and the United Kingdom differ in their definitions of "urban areas?" Why do you think this is the case?

3. How can urbanization be used as an economic development indicator in various parts of the world?

4. Which continental region has the most cities with populations of 10 million or more? Why do you think this particular part of the world has experienced such tremendous urban population growth rates?

5. Urban location is usually tied to the original function of the settlement such as defense, trade, resources, administration or religion. List an example for each from your readings. How would you classify the city where your college or university is located and why?

6. Where and what is the "City of the Dead" and how did it earn this unusual title?

PART II: Gaining Your Census: The Rise and Fall of U.S. Urban Populations

For this activity you will need to access the United States Census Bureau web site and read and article and review some tables relating to urban population change. To access this site, go to a map of the *United States* and click on *Web Links*. Select "*United States Census Bureau*." Use the *search* tool on the web page and enter "*suburbs*" then click on the article entitled "*Population of the 100 Largest cities and Other Urban Places in the United States: 1790 to 1990.*" Read the relatively short narrative (you don't have to go through all the explanations of the tables at the end of the article), and then download the information for Table 1, "*Rank by Population of the 100 Largest Urban Places, Listed Alphabetically by State: 1790-1990.*" You do not need to download any additional tables (Helpful Hint: If you print out Table 1, change your printer paper orientation to "*landscape*" instead of "*portrait*"). Now you are ready to answer these questions.

7. Beginning in 1930, how did the U.S. Census define incorporated "urban places," in terms of population size?

8. Take a moment to review Table A, "Population, Land Area, and Density for the 20 Largest Cities: 1990." Which U.S. city has the most population? Which is largest in land area? Which city is the most densely populated?

9. Now look at Table B, "Population of the 20 Largest Cities and Urbanized Areas: 1990." When you take into account suburban populations, the rank of cities begin to change. Which cities that rank in the top ten in "Urbanized Area" population don't have similar ranking in "City" column. Can you make an educated guess as to why?

10. The next two questions refer to Table 1 which ranks the 100 largest urban areas in the United States between 1790 and 1990 so keep it handy. We already know what the top ten cities were in 1990, but what about 1790? List the top ten cities by population size below for 1790. How do those compare to the ranking from Table B?

11. Take a look at statistics for the state in which you live (which may not be where you are attending college). Which cities made it to the list for this 200-year period? How would you describe the trends in rankings---i.e. did most grow or did most decline, or did some completely drop out of the list? Based on what you know about the cities in your state, what factors account for the changes in rank?

Keywords: **cities, urbanization, suburbs, U.S. Census, population**

Chapter 11: Do Orange and Green Clash?: Residential Segregation in Northern Ireland

I'll See Your Bernadette Devlin and Raise You an Ian Paisley

Most of you reading this weren't even born with the "Troubles" began in Northern Ireland (a.k.a. Ulster) in the 1960s. Nor does the title of the chapter activity mean much to you. Bernadette Devlin was a young firebrand member of parliament (MP) elected from a heavily Catholic section of Northern Ireland. Her nemesis was Ian Paisley, also an MP and a staunch Protestant loyalist. Northern Ireland is the smaller northern end of the island with the same name, a part of the larger United Kingdom that also includes England, Wales and Scotland. The larger southern portion of the island on which Northern Ireland is located contains the Irish Free State (a.k.a. Eire), a country that is wholly independent of the United Kingdom. Most of the residents of Ireland are Roman Catholic. Most of the population of Northern Ireland owes its allegiance to the Church of England (e.g., Episcopalians as we call them in the United States). There are, however, significant Catholic minorities in the major cities of Ulster especially Belfast and Londonderry and along the southern border region near Eire.

Why, when Episcopalians and Roman Catholics in the United States are entering into dialogue about sharing the Eucharist and other forms of shared worship experiences in the spirit of Christian ecumenicalism, are Protestants and Catholics in Northern Ireland still shooting at each other? Good question. The simple answer is that they soon may not be if all parties agree to the Peace Accord recently brokered by President Clinton working with Tony Blair, the British Prime Minister and representatives from Ireland and even the political arm of the Irish Republican Army, *Sinn Féin*. But truces and partition lines have been in place in Northern Ireland since 1969. Catholics in Northern Ireland detest the presence of British troops in their neighborhoods ostensibly to keep the peace and Protestants certainly don't like the terrorist tactics of the Irish Republican Army (IRA). The IRA has disrupted commerce, killed people with car bombs and generally wreaked havoc on the entire United Kingdom, not just Northern Ireland.

As with many arguments, the fight is over more than just religion. Catholics feel that are treated as second-class citizens in Northern Ireland. They claim that they are always the last to be hired and the first to be fired from any job. The geographer Frederick Boal has written about the Fall Road-Shankill divide in Belfast, Northern Ireland. This divide was designated as a barricaded truce line after 1969 and the open hostilities seemed to die down with the imposition of British troops sent to enforce the peace. South of the divide is the Catholic neighborhood of Clonard. To the north is the heavily Protestant neighborhood of Shankill. The segregation indices discussed in the chapter would certainly be able to demarcate this strong distinction. The Shankill neighborhood always displayed orange bunting and painted the curbs orange on July 12th. Using *Encarta's Virtual Globe '99*, let's find out why. Specify '*Northern Ireland*' in the *Find* area (upper left). Now, click on the sidebar marked *Sights and Sounds* and click onto the slide entitled "*Children at an Orange March*."

1. After reading the caption, describe in your own words the origins of the animosity of the Protestants for the Catholics and vice versa.

An interesting *Web Link* is the one to the *Irish News*. This newspaper, published in Northern Ireland, is very pro-Catholic. Protestants, according to Boal's study, read the *Belfast Times* almost exclusively and Catholics read the *Irish News*. As of this time of this writing, the peace plan proposed by Prime Minister Tony Blair and endorsed by most, but not all, Catholic and Protestant groups tops all other news and may be the best chance yet for a lasting peace.

2. But time marches on. What are the featured stories in the issue of the *Irish News* that the web site has brought up?

Now let's switch to a focus on the Republic of Ireland. Using the *Find* command type in 'Ireland'. The textual material in the sidebar labeled *Society* provides good background to the current situation. Click on that and then scroll down or click on the section called *Recent Decades* and read it carefully and then answer the following questions.

3. What does the Gaelic phrase *Sinn Féin* mean?

4. How long has Northern Ireland been separate from Ireland?

5. How would most Irish citizens like to see the issue of Northern Ireland resolved?

6. How do you personally think this festering debate, exacerbated by residential segregation, mutual distrust and misunderstanding will eventually resolve itself?

Keep your eyes open to the newspaper and follow the story closely. The last time an accord was almost reached, an act of IRA terrorism scotched the deal. Let's hope that for the sake of world peace this current accord may finally end the longstanding "Troubles" in Northern Ireland.

Keywords: **Northern Ireland, Ireland, Roman Catholic, Protestant, residential segregation, Sinn Féin, Irish Republican Army**

Chapter 12: The Rise of Nationalism and the Fall of Yugoslavia
Timor or Less: Indonesia's Claim at Stake

Have you ever heard of *Timor*? Did you even know it was an island let alone where it might be? Does it seem that some places on earth are so far removed from you that what happens there has little or no impact on your life? Such is the case with Timor, an island that is divided into East and West and claimed by Indonesia. But, the residents of East Timor recently voted their portion of the island to be independent of Indonesia by an overwhelming margin (almost 80 percent of the electorate favoring independence). The election set off a period of murderous violence against that majority electorate. Until the intervention of a UN peacekeeping force led by Australia, the island was wracked by violence of guerrilla groups loyal to Indonesia and, by their complicity and failure to stop the carnage, Indonesian troops who were ostensibly there to keep the peace. At the time of this writing, East Timor is very much in the news. We need to go back to the beginning to find out what is behind the desire for independence and why the world finally decided to intervene to resolve the issue of East Timor's independence once and for all.

Go to map of *Timor* using the *Find t*ool. Read a short description of the island by clicking on *Geography*.

1. How is the settlement history of Timor reflected in its ethnic composition?

2. How did the withdrawal of colonial powers make Timor vulnerable to exploitation from Indonesia?

3. What role has the United Nations played in resolving human rights issues in East Timor?

Part II. Let's look further into the conflicts between Timor and Indonesia by going to *Web Links* and clicking on "*East Timor Human Rights Centre.* The Internet address has been changed but the new URL (address) is given. *Click* on it and the file will open. Based on your review of the reports and media releases with which the site concerns itself, answer the following questions.

4. Where is the East Timor Human Rights Centre located?

5. Is there a significant ex-patriate community of East Timorese in Australia?

6. How would describe the character of the government of Indonesia prior to the vote on East Timor's independence?

7. What role did the military play in the government of Indonesia? Of East Timor?

8. Despite a rapidly expanding economy, why do social and economic disparities remain such a chronic problem in Indonesia?

9. What examples of human rights abuses are cited in these articles? Do you see a common thread between them?

10. What evidence was there of persecution in East Timor during the period of Indonesian occupation based on the residents' ethnicity and religion?

11. Who is Bishop Belo and what role has he played in the East Timor conflict?

12. How would you describe the status of women in East Timor? Why are they reluctant to report that they have been victims of crimes? How would you describe the crimes of sexual assault against women in East Timor and how is the Indonesian military involved in these crimes?

After reading about conditions in Yugoslavia in Chapter 12 of your textbook, do you see any similarities or differences between the situation in southeastern Europe and Indonesia? Which region do you believe will be the first to find peace?

13. What role has geography played in the political fortunes of these areas?

14. What do you think will have to happen before students like yourselves become more concerned about Yugoslavia and Indonesia? East Timor?

Do you think your friends could even find these places on a map? Could you?

Keywords: **Indonesia, Timor, East Timor, human rights, gender, political geography**

Chapter 13: Human Impact on the Environment
How Would You Like Your Animals—Rare?

Think of yourself as the zookeeper to the word, responsible for the welfare of everything from aardvarks to zebras. It would not take you long to figure out there is a colossal variety of animals on the planet, each with their own unique habitats, food requirements, and frailties. And as a good zoo manager, you would want to do an inventory of the earth's animals, find out where and how they live, and what if any dangers they may be facing. In your research you would also learn that many creatures are facing incredible odds competing with humans for resources and space and in virtually every case the humans are winning the war. In this activity you will have an opportunity to study the crises faced by the world's wildlife. Let's use *Encarta Virtual Globe '99* to zoo-m in on the issues.

Lions and Tigers and Bears, Oh My!
The first thing you need to do to evaluate the state of the world's zoo, is find out where the animals live and which species are the most threatened. To accomplish this, go to a map of the *world*, click on *Features*, then *Statistical Center*. When the statistical table appears on your screen, select the "*Environment*" category. In the right hand column you should see a list, beginning with "*Amphibian species, known*." For each of the following categories, list the countries that are ranked in the top five. As you go through each group you will want to see the data in tabluar form so click on the *Statistic Table* button at the bottom of the screen. After you complete you lists, answer the questions that follow.

- Amphibian Species, Known

- Amphibian Species, Threatened

- Bird Species, Known

- Bird Species, Threatened

- Fish Species, Known

- Fish Species, Threatened

- Forest, Annual Rate of Deforestation

- Land, Area Protected as a Share of the Total Land

- Mammal Species, Known

- Mammal Species, Threatened

- Reptile Species, Known

- Reptile Species, Threatened

1. What are the only categories of threatened species in which the United States does not appear? Why do you think the U.S. may be doing better than other countries with protecting these animals?

2. In which category is the U.S. ranked the highest in threatened species---and by a large margin? Why would the U.S. be such an unfavorable environment for the survival of these animals?

3. What pattern do you see emerging as you compare threatened and known species with regard to their respective world economic development regions?

4. Go back to the *Environment* statistical table and select the "*Threatened species.*" List the top ten countries for numbers of threatened species. Based on your knowledge of the world and any information you may find in your text or in *Encarta Virtual Globe '99*, why do you think these countries show up so high on this list? What do most of these countries have in common that would create such strong competition for land and resources between humans and animals?

Keywords: **environment, endangered species, economic development**

Physical Geography
Exercise 1: The Earth as a Rotating Planet
Zoning in on Greenwich:
Giving You the Time of Day

It's time to go to Greenwich, England, and zone in on the history and applications of time zones. So, let's get started. Using the *Find* box, create a map of *Greenwich*. Take a few minutes and look at the information about Greenwich contained in *Geography* and *Sights and Sounds*. Once you have this background information you are ready to do some "time" traveling via the Greenwich *Web Links*.

PART I: Greenwich: Time on Their Hands
Go to the *Greenwich 2000* web site. Scroll down to the very bottom of the page and click on *FAQ* (i.e., frequently asked questions) under the heading *Greenwich Time*. Read through this section then answer the following questions.

1. What transportation innovation necessitated the creation of time zones?

2. Why was Greenwich chosen as the center for world time keeping?

3. What is the name of the "line" that passes through this British town?

4. What accounts for the differences between "UT1" and "UTC?"

5. Why did UTC replace "GMT?"

6. What is "atomic time" and how does it relate to the earth's rotation?

7. Astronomers in Greenwich were "dropping the ball" long before New Year's Eve revelers in Times Square got into the act. Why did Greenwich scientists initiate this tradition in 1833? Why did they drop their "time-ball" everyday at 1:00pm?

8. What is British Summer Time (BST) and speculate why it was introduced?

PART II: Welcome to the Time Zone

While you still have the *Greenwich 2000* web site handy, go back to the home page and in the same box where you found the *FAQ*, click on *Zones*, and consider the following questions.

9.   Why do you think the military uses different time zone designations than civilians?

10. Where would you be if you lived in the "Whiskey" time zone?

11. What location is associated with the "Bravo" time zone?

12. What are the military designations for time zones in Russia?

And speaking of Russia, take a ride on the Trans-Siberian railroad and try out your time zone expertise. Using the *Find* tool, create a map of *Russia*. Go to *Sights and Sounds* and click on *"Riding the Trans-Siberian Railway."*

13. How man times zones would you cross if you took a ride on the Trans-Siberian railroad heading east from Moscow to the end of the line at Vladisvostok?

14. If you were boarding your train Moscow at 5:00am, what time would it be in Vladisvostok? (Hint: To verify your answer, you can go back the *"Greenwich 2000"* web page and click on *"Map of World Times"* or consult the time zone map in your textbook.)

Keywords: **time zones, Greenwich, atomic time, Russia, Trans-Siberian Railroad**

Physical Geography
Exercise 2: The Global Energy System
Holy Ozone Batman!

Have you ever purchased a pair of sunglasses or a bottle of sunscreen lotion for the purpose of ultraviolet (UV) ray protection? Imagine how much lotion it would take to protect an entire continent! Recent scientific studies have indicated that the ozone layer above Antarctica could use a lot of sunscreen. Antarctica is a living laboratory for studying the ozone layer, the atmospheric shield that protects the earth from harmful UV rays. In this activity, you will explore Antarctica to learn more about the continent's environment and how the health of the region's ozone layer may affect the entire world.

PART I: Frozen Assets: The Antarctic Environment
To obtain some vital information about *Antarctica*, go to a map of the region using the *Find* box and select on *Geography* and *Sights and Sounds*. Be sure to click on "*ozone layer*" in the final paragraph of *Geography*. When you are done reading and reviewing the material, answer the following questions.

1. What are the only land animals and invertebrates that inhabit Antarctica?

2. How is it that Antarctica could be classified as a "true desert?"

3. As a continent, Antarctica has three distinctions that place it in the record books---what are they?

4. What is the geologic origin of the Transantarctic and the Ellsworth Mountains?

5. What percent of the world's fresh water is bound up in Antarctic ice?

6. What happens to the size of Antarctica during the winter months and why?

7. Which countries have made territorial claims to Antarctica? Why would these nations be interested in staking a claim on this frozen wasteland?

8. In which atmospheric layer is the ozone layer located? At what altitude would you find the ozone layer?

9. How is ozone formed and how does pollution affect the ozone layer?

10. Why should humans be concerned about holes in the ozone layer?

11. What are some of the concerns that scientists have about developing tourism in Antarctica? Would you ever like to visit this place---why or why not?

PART II: And You Thought NOAA Was Only Interested in Arks?

For this section, you will need to access the *Web Links* for *Antarctica*. Click on *Web Links* and select "*The New South Polar Times*." When the homepage for this link appears, click on "*Questions and Answers,*" which will take you to the link for *"NOAA's Role in the South Pole."* When you get to the NOAA page, pay particular attention to the final section, "*Why is all this important?*" Read through each section and consider the following questions.

12. What are the five things *other than ozone* that NOAA measures in the Antarctic atmosphere?

13. What does a Dobson spectrophotometer measure?

14. What makes the Antarctic a good place to conduct experiments for ozone and these other atmospheric variables?

15. How does the "heat sink" associated with Antarctica permit researchers to study early warning signals for world climate change?

PART III: THE HOT OZONE

Now we will compare the ozone issues of Antarctica to a much warmer and wetter environment, Brazil. Go to a map of *Brazil* using the *Find* tool and select *Land and Climate*. Read the "*Environmental Issues*" section and answer the following questions.

16. What is the difference between the "heat sink" of Antarctica and the "carbon sink" of the Amazon?

17. Antarctica had several environmental superlatives associated with it---what about the Amazon? What are some of the unique characteristics of Brazil?

18. How is the loss of rainforest linked to global warming?

19. How does urbanization in Brazil affect the conditions of the earth's ozone layer?

20. What is Brazil doing to correct the damage to their environment and protect it for future generations?

Keywords: **Antarctica, ozone layer, environment, NOAA, Brazil, rainforest**

Physical Geography
Exercise 3: Air Temperatures and Temperature Cycles
Keeping Your Head Above Water:
Sea Levels and Global Warming

Has anyone ever offered to sell you oceanfront property in Arizona? Even with the most dire predictions for sea level rise due to global warming, Phoenix residents shouldn't start worrying-- yet. Many scientists believe that a general warming of the earth's temperatures and the subsequent warming and expansion of ocean waters could result in as much as a 35 inch rise in sea levels. While three extra feet of water depth may not sound like much to you, for residents of low-lying or flood-prone areas such as the Maldives or Bangladesh, the results would be devastating. In this activity you will examine some of the predicted impacts of global warming as they relate to sea level and environmental change and take virtual field research trips to the Maldives, Bangladesh, and Canada.

PART I: Islands No More?
We begin our exploration with a trip to the *Maldives*. Using the *Find* box, create a map of the island nation. Click on *Land and Climate* to learn about the physical environment of the Maldives. Be sure to check out the slide titled, "*Coral Islands of the Maldives,*" at the beginning of *Land and Climate* and answer the following.

1. How many feet would the Indian Ocean sea level have to rise to place 80% of the Maldives under water?

2. If you were a resident of the Maldives, what would be your biggest concern about water issues in the present? What has caused this water resource problem?

3. Let's say the sea level does rise and inundates the islands. Using the *Measuring Tool*, where would be the closest landfall for persons escaping the Maldives and how many miles would it be to shore?

PART II: Bangladesh: Water, Water, Everywhere
Bangladesh, like the Maldives, is particularly at risk should global warming result in higher sea levels. Using the same approach as for the Maldives, create a map for *Bangladesh* using the *Find* option and review *Land and Climate*. Be sure to click on the slide "*Nation of Rivers*" at the beginning of the text and then answer the following questions.

4. What two things do Bangladesh and the Maldives have in common with respect to elevation and fresh water supplies?

5. What is the capital of Bangladesh and what is its approximate elevation (check the Map Legend to determine this). Would this city be under water if sea level rose three feet? If you were going to move the capital to higher ground, which city in Bangladesh would you choose and why?

PART III: Canada Dry or Canada Wet?

Our final stop on our global warming tour is *Canada*. Far from the tropical locations we have previously studied, Canada faces a different set of problems that could result from global warming. To learn about how climate change could affect Canadians, create a map of Canada using the *Find* box. Next, select *Web Links* and go to the *"Environment Canada"* homepage (choose either the English or French version) and enter *"The Green Lane."* Click on *"Issues and Topics"* and select the *"Climate Change"* link. From the left side of the screen, first choose, *"What is Climate Change?"* After you are through reading this section, go to the *"Factors Affecting Global Climate"* and *"Climate Trends"* links and then to, *"How Will Climate Affect You?"* By now you should be an expert on global warming issues in Canada and well prepared to answer the following questions.

6. According to the United Nations Framework Convention on Climate Change, what is the cause of climate change?

7. How much warmer is earth today than it was 100 years ago? Have there been any periods of cooling in the same time period---if so when?

8. In the *"Climate Trends"* discussion, it is stated that nighttime temperatures over land have increased more than temperatures during daylight hours. Why do you think this is the case?

9. How much warmer was Canada in 1992 than 1895? What are some human and natural factors that have contributed to the general increase in Canadian temperatures?

10. How do atmospheric aerosols contribute to global warming?

11. How would global warming affect urban populations in cities like Toronto and Montreal?

12. How would Canadian agriculture be hampered by increased air temperatures? Could global warming actually help farmers--how?

13. What are scientists predicting for the Great Lakes should global warming prove to be a serious problem?

14. How would global warming affect the Atlantic provinces of Canada and is there any evidence that this is already happening?

Keywords: **global warming, sea level rise, Maldives, Bangladesh, Canada**

Physical Geography
Exercise 4: Atmospheric Moisture and Precipitation
Somewhere Over the Rain Shadow

Some mountain ranges are so tall that they can actually block weather patterns, thus modifying the weather and climate on different sides of the mountains. In the United States, mountain ranges that have sufficient elevations to dramatically affect weather and rainfall patterns would include the Sierra Nevada Range in California and the Cascade Range in Oregon and Washington.

As we have learned, all of the temperate mid-latitude areas in the United States are subject to the Prevailing Westerlies (i.e. the consistent air flow from west to east across the mid-continent). The sheer mass and height of these large Western mountain ranges force east flowing air masses aloft. As these air masses rise, they cool in temperature and begin to condense. If conditions are right, there is considerable precipitation on the windward side of the mountains. This phenomenon is called orographic precipitation and in higher elevations or in winter, this may fall as snow. A large amount of snow pack in the winter and a rapid rate of melting in the spring can often determine if sites on the windward side of the mountain may be in danger of flooding. The leeward side of the mountain (i.e., away from the wind) is less likely to receive much precipitation as the air mass sinks which creates the rain shadow effect.

PART I: Orographic Precipitation: An Uplifting Experience
Let's take a closer look at some examples of how landforms affect precipitation patterns. First, go to the *Find* box and search for "*orographic*" and when that word appears under *Content*, click on that selection and read more about this precipitation pattern.

1. Where do you think the rainfall would be greatest—in the more southerly Sierras or the more northerly Cascades? Why?

2. What are the relative temperatures of the water bodies over which the air masses that dominate the Sierras and the Cascades respectively are formed?

3. In general, what is the moisture holding capacity of a warm air mass vis-à-vis a cold air mass?

PART II: Where is the Virtual Airsickness Bag?: Have No Fear of Flying
Now click on *Virtual Flights* under the heading at the top of Encarta marked *Features*. This feature takes you to three-dimensional representations of a variety of landscapes. If you have not yet installed the shareware program entitled *Macromedia Shockwave*, do so when the computer instructs you to. This will take some time (about nine minutes to download on a 300 MHz machine with an Ethernet connection), but once downloaded and installed (which requires the insertion of Disc 2), the sky is the limit on the number of Virtual Flights you might wish to take in the future. Remember: installing and running this portion of the activity will require BOTH *Encarta Virtual Globe '99* CDs—Disc 1 and Disc 2 (Installation and Resources). With the *Macromedia Shockwave* software installed, click on *North America*. The screen will start with the west coast of California. Experiment with the distance and directional controls that guide the "flight" and head north up the coast to the Puget Sound lowland and the (emerald) city of Seattle. Experiment with stopping the plane in flight (something that is impossible in the real thing), dip

down to a lower altitude and then zoom back up to a reasonable cruising altitude. Once you reach Seattle, take a virtual traverse from western Washington to the city of Spokane in eastern Washington (note: a camera angle of –25 degrees seems ideal for picking up on subtle elevation changes as you make the traverse). Now, answer the following questions:

4.   How would you describe the landscape changes you see there?

5.   What are the names of the two peaks that are visible from Seattle?

6.   Do you think that they are volcanic in origin? Why or why not?

Now, using the *Find* command, bring up a map of the state of '*Washington*'. The annual precipitation in the state can vary from more than 150 inches of rain in the temperate rainforest of the Olympic National Park to less than 20 inches of naturally occurring rainfall in the rain shadow of the Cascades. Despite this low amount of naturally occurring rainfall, Washington's apple valleys (e.g., Yakima, Wenatchee), produce the famous red and yellow delicious apples that are shipped everywhere in the United States.

7.   Where does the water for these abundant apple orchards come from?

Click on *Sights and Sounds* and examine the slides entitled '*Majestic Mount Rainier*' and '*Hoh Rain Forest*' and answer as best you can the following questions:

8.   How extensive is Mount Rainier? How many glaciers does it contain?

9.   The Hoh Rain Forest is in the Hall of Mosses in Olympic National Park. From an examination of the map of Washington state, explain as effectively as you can why the National Park receives over 150 inches of rainfall a year and the eastern flank of Mount Rainier less than 20.

Keywords: orographic effect, Prevailing Westerlies, Cascades, Sierras, rain shadow

Exercise 5: Winds and the Global Circulation System
A World Wind Tour

The age of sailing vessels and exploration of unknown territories was literally fueled by the wind. Knowing where and when the winds would blow meant the difference between success and failure, life and death. It's no wonder that so many places across the globe gained their prominence or even their names due to their location in relationship to the wind. Let's go with the flow and explore the world by the winds on the good ship Encarta. We will visit four locations to study the historical impacts of trade winds, Zanzibar Island, Australia, Fiji, Hawaii, and Ecuador.

PART I: Zanzibar None
Before we being our world wind tour, you should read more about trade winds. To do so, enter "*trade winds*" in the *Find* box and review the material found in the *Glossary* (or you can simply click on "*trade winds*" as they are highlighted in the *Geography* description of Zanzibar Island). Now we are ready to begin our journey on Zanzibar Island. Create a map of this Tanzanian territory using the *Find* box and then click on *Geography* and answer the following.

1. According to the trade winds map in the Glossary, what is the prevailing wind direction that would affect Zanzibar Island?

2. According to legend, how did Persian sailors travel to and from Zanzibar?

3. Which colonial powers, famous for their sailing vessels, controlled Zanzibar Island since the 16th century?

4. How did Zanzibar's location fit within slave trading routes of the late 18th and early 19th centuries?

PART II: Take a Bight of Australia
As we leave Zanzibar Island and head east to Australia, we will witness a number of environmental conditions that are shaped by prevailing wind patterns. To take a closer look, create a map of *Australia* using the *Find* tool. Read the descriptions of the Australia's environmental conditions by clicking on *Land and Climate*. Be sure to click on "*Great Australian Bight*" which is highlighted under the "*Location*" heading at the beginning of the article. When you get to the Bight, take a look at a coastal scene via *Sights and Sounds* as well as read the short description of the region by selecting *Geography*. With all this information at hand, you should be ready to answer the following questions.

5. What happened to many species of Australian wildlife after sailing vessels brought colonizers to the continent? What types of animals were introduced to Australia as a result of colonization and what have been the environmental consequences of adding these new species to the continent?

6. What type of climate dominates the northern portion of Australia but loses its influence in the south?

7. From what direction do the prevailing winds flow across Australia? How does this affect the precipitation patterns?

8. What sort of wind patterns affect climate patterns in Tasmania? When is their rainy season?

9. What has been the major cause of desertification in Australia?

10. Why is Australia a good candidate for wind power development?

PART III: Getting S'Pacific About Wind Patterns

Our next destinations include a series from Fiji, Hawaii, and Ecuador. For each we look at how sailing vessels and wind have shaped the history of these places. First we take landfall in Fiji. Using the *Find* box, create a map of Fiji's largest island by entering "*Viti Levu.*" Select *Geography* and read the brief description of how Viti Levu's location in the Pacific affects their climate, vegetation and agriculture. Next, sail on to Hawaii and visit Kauai Island. Take a look at Wailua Falls in *Sights and Sounds*. Your final stop on your Pacific tour is Ecuador. Using the *Find* box, create a map of "*Guayaquil*" and read the *Geography* section. Be sure to click on "*Humboldt Current*" while you are reading about Guayaquil. Now you are read to tackle the following questions.

11. Which prevailing wind patterns affect Fiji? Hawaii? Ecuador?

12. How do trade winds help to create large rainforests on Viti Levu?

13. Why do you think the northeastern part of Viti Levu is drier than the southeastern part? What crop is grown in the northeast?

14. What and where is the "wettest place on earth?"

15. How do trade winds affect the precipitation regime of Kauai Island?

16. What is the Humboldt current and do the winds associated with it affect Guayaquil?

17. What happens to the climate of Guayaquil from January through April and why?

Keywords: **winds, trade winds, sailing vessels, historical geography, climate, Zanzibar Island, Australia, Fiji, Hawaii, Ecuador**

Exercise 6: Weather Systems
Going, Going, Gone with the Wind

Tornadoes and hurricanes can unleash some of the most lethal damage of any natural hazard event. The winds in a tornado are so strong that they can drive a straw through a telephone pole. It is almost impossible to measure their maximum speed but it is well over 300 mph. The destruction they usually leave in their wake is somewhat localized and erratic, but the rare tornado can cut a swath through an area of a mile in width and many miles in length. A hurricane, on the other hand, is a dangerous storm over a much larger area than a tornado. Wind speeds in the most damaging hurricanes (Category 5) can approach two hundred miles per hour. The area affected by hurricanes can experience high winds, storm surges, flooding, heavy rain, and even tornadoes spawned by the main storm itself.

It is estimated that 54 percent of the U.S. population lives within 50 miles of a coastline so a large proportion of the population could conceivably be at risk from hurricanes. Let's find out more about these widespread and sometimes devastating cyclonic storms. First we will review some information about hurricanes in general and then focus on Hurricane Andrew, the 1992 storm that brought great devastation from Florida to Louisiana.

PART I: Counterclockwise Swirls
To begin our background research on tornadoes and hurricanes, enter each term in the *Find* box and read the *Glossary* descriptions for each. Be sure to click on "*Monsoons, Tropical Storms, and Tornadoes*" under *Related Topics*.

1. Major tropical storms in the Atlantic and western Pacific Oceans are known as hurricanes. What are they called in the Indian Ocean and in Australia?

2. According to the *Glossary* map, when is western Mexico at highest risk for hurricanes?

3. How far does the risk zone for cyclones extend from December through March?

4. Where do tornadoes most commonly form and why?

5. Why is it so difficult to accurately predict where and when tornadoes will strike?

6. Where do you find the most prominent monsoon climate and why?

7. What are the word origins for hurricanes and typhoons?

8. In the Southern Hemisphere, in which direction do cyclones rotate?

9. How do trade winds affect tropical storms?

10. What are the "breeding" conditions for tornadoes?

11. What is the name for a tornado that passes over water and why do you think these storms are weaker than those that move across land?

PART II: Keeping An Eye on the Storm

In this section we will take a closer look at hurricanes via a web site that does not have a direct link to Encarta Virtual Globe. To access this site, enter the following address in the box where your browser usually lists the URL you are using:

http://weather.unisys.com/hurricane/atlantic/index.html This will take you to the Unisys web page for "*Atlantic Tropical Storm Tracking by year*". The index that is provided covers hurricane data for the Atlantic from 1886 to 1998. Also included is the Saffir-Simpson Scale code for the maps you will be looking at for this activity. The Saffir-Simpson Scale is used to classify hurricanes by their strength. You will need to do several things from this index page. First, click on "*1992*" and take a general look at the storm patterns for that year, in particular, the "*tracking information*" for *Hurricane Andrew*. Next, go back to the index page and select "*1998*" and review the general tracks for all storms but in particular the tracking information for *Hurricane Mitch*. Once you have all that information at hand (if possible, you may find it helpful to print these pages out for easier reference), you are ready to answer the following questions.

12. Take a look at the "Individual Storm Summary" for both Hurricane Andrew and Mitch. How many days did each storm last, what were the maximum sustained winds, and the lowest barometric pressure readings for each?

13. When you compare the maps for all hurricanes and tropical cyclones in 1992 and 1998, what are some of the most striking differences in their numbers, patterns and paths?

14. How many days did each storm remain a "Category 5" hurricane?

15. With the "tracking information" tables handy, note the geographic grid coordinates (latitude and longitude) for each location that reported Category 5 conditions for both Andrew and Hurricane Mitch. Using the Dynamic Sensor tool, find the locations where each storm was identified as a deadly Category 5 storm and list them below.

104

16. Both Andrew and Mitch proved to deadly storms and highly destructive. But if Hurricane Andrew was not a Category 5 storm for very long, why do you think it caused such great damage in South Florida and later in Louisiana?

17. Compared to Hurricane Andrew, Mitch was a much deadlier storm. Over 10,000 people lost their lives in Honduras, Nicaragua and surrounding areas due to Hurricane Mitch. While fewer than 30 persons lost their lives in Florida from Hurricane Andrew, the storm is considered to be the most expensive natural disaster in U.S. history to date causing over $25 billion in damage. When you compare these two storms, why do you think Mitch wrought such horrific loss of life, more so than Hurricane Andrew? (Hint: If you think you need more information about Honduras, go to a map of that country using the *Find* tool and read about it's geography, economy, and social conditions.)

Keywords: **tornadoes, hurricanes, typhoons, tropical storms, cyclones, Saffir-Simpson Scale, Hurricane Andrew, Hurricane Mitch, Florida, Honduras**

Exercise 7: The Global Scope of Climate
Köppen, Schmerpen: What Can Botany Tell Us About Climate?

Of all the climate classification systems, the one that is still the most widely accepted is that first developed in 1918 by the Austrian plant geographer and amateur climatologist, Vladimir Köppen. The system he developed has been considerably modified by a succession of scientists. There has certainly been a substantial increase in our knowledge of climates in areas of the world that were inaccessible or not studied by Köppen and his contemporaries. Köppen was a geographer interested in the ecological niches of different types of plants and their tolerance for varying combinations of temperature and moisture conditions. He was in search of plant species that might define the boundaries between different climatic regimes. These would be plants that are sensitive enough to thrive in one type of climate but languish in another. One such plant was the citrus we know as the key lime. Have you ever eaten key lime pie? What color was it? If it was truly made from key limes, the color would have been a yellowy-green, not the much deeper green we associate with the limes commonly found in grocery stores outside of south Florida. The key lime needs the megathermal climate (A or B in the Köppen classification scheme). They also need adequate moisture and a B climate (without some type of supplemental irrigation) would not provide the needed moisture. An examination of the Köppen map of world climates reveals only one location in the U.S. that has an A climate (an Aw savanna climate to be exact). Where would this be? Why...the Florida Keys. Thus the name "key" lime.

In this activity we will travel the world testing Vladimir Köppen's climate system and see how it might work. Because the Köppen classification system is so intimately tied with vegetative growth, the climate types may bear more resemblance to *Encarta Virtual Globe '99s* "*ecoregions*" than its climate region maps. So keep your textbook handy for reference as we traipse across the globe in search of climate clues.

PART I: Club Mediterranean Climate
To begin studying the connections between vegetation and climate go to *Features* and select "*Global Themes.*" When you get to the screen that lists a variety of subjects, select "*The Living World*" and then "*Mediterranean Woodlands.*" Take a few extra minutes and click on each of the images included in the article as well as the *Ecoregions* map.

1.   Other than around the Mediterranean Sea region itself, where else do you find Mediterranean climates?

2.   What are sclerophylls and how do they survive drought conditions?

3.   What role does fire play in Mediterranean woodland ecosystems?

4.   How are Mediterranean woodlands in South Africa different from those in Italy and California?

5.   Goats are notorious for eating anything and everything, however, some Mediterranean woodland species are efficient goat repellents. What types of plants are these and how do they "get their goat?"

PART II: Get Boreal

Return to "*Global Themes*" via *Features* and select "*The Living Planet*" and then "*Boreal Forest Regions.*" Read though the description of this climate type, look at the *Ecoregions* map as well as the three images included in the article.

6. Compare the *Ecoregions* map to the Köppen climate map in your text. What would be the Köppen climate designation for this environment?

7. What types of trees are commonly found in boreal forests and how have they adapted to the severe climate conditions?

8. Why are there so many ponds in taiga regions?

9. Why are boreal forests in North America more diverse in species than the taiga of Eurasia?

10. Why are spruces and firs better adapted to boreal climates than deciduous trees?

11. Why do you think trees become shorter and more widely spaced as you move north out of boreal climates and toward tundra environments?

12. The boreal forests of Slovenia, Appalachia, the Sierra Nevada, and Rocky Mountains are well south of the normal extent of boreal climates. Why are these locations are then classified as boreal?

PART III: Three Prairie Dog Night: Grasslands

Our final stop at the salad bar of climates is the world's grasslands. Return to, "*Global Themes*" and "*The Living Planet*," and this time select "*Temperate Grasslands*." Once again, be sure to look at the *Ecoregions* map and at each of the six images that accompany the text and then answer the following.

13. Using the map *Legend*, determine which of Köppen's climate classifications best describes the temperate grassland regions of North America.

14. How are grasslands differentiated from deserts and forests?

15. Other than precipitation and temperature, which two factors determine the formation of grasslands?

16. Obviously grasses are the dominant plant type in grasslands, but what are some of the other types of vegetation you might find and where might you find them?

17. What accounts for the differences between tall grass and short grass prairies?

18. How do the grasslands of Eurasia compare to those of North America?

Keywords: **Köppen, climate, ecoregions, Mediterranean climate, boreal forests, grasslands, prairie**

Physical Geography

Exercise 8: Earth Materials and the Cycles of Rock Change
Don't Take All Rocks for Granite

Earth is probably the mother of all recyclers. It starts out with igneous rocks, breaks them down and reconfigures them into sedimentary rocks. Then, through recrystalization--voilà—metamorphic rocks. And then you can melt them all down again and start all over. In this activity you will take a magical mineral tour of the world and visit sites sampling all three rock types--igneous, sedimentary, and metamorphic.

PART I: Northern Ireland: A Belleek Landscape

Your first stop on your rock tour is Northern Ireland. Create a map of this part of the *United Kingdom* using the *Find* box. Click on *Geography* and read about the area's physical environment. After reviewing the narrative, switch over to *Sights and Sounds* and view the slides titled, "*Newcastle on Dundrum Bay*," and "*Giant's Causeway*."

1. What types of minerals are commercially exploited in Northern Ireland? Classify each by their origin—igneous, sedimentary, or metamorphic.

2. What type of rock would you find on the peaks and slopes of the Mourne Mountains. Which rock classification would this be?

3. What is the "Giant's Causeway?" What does this formation tell you about the geologic origins of the island?

PART II: To Eire Is Human

The second stop on our rockumentary is the *Republic of Ireland*. Go to map of this country by using the *Find* option and take a look at *Sights and Sounds*. Review the slides entitled, "*Ireland's Green County Kerry*," *Ireland's Aran Islands*," and *'Rock Garden of the Burren*."

4. Which of the three rock classifications are represented in all three slides taken in Ireland? What kinds of this rock type are illustrated?

5. What are Macgillicuddy's Reeks and what type of rock would you find there?

6. How has limestone become as much as part of the cultural landscape of the Aran Islands as that of the natural landscape?

7. Where would you find a "limestone desert" in Ireland? How has this landscape been described and what is its primary use?

109

PART III: I Never Metamorphic I Didn't Like

The final leg of our rock tour include stops in *Australia* and *Italy*. First, create a map of *Katherine Gorge, Australia*, using the *Find* box. Click on *Sights and Sounds* and view the side of the gorge—isn't it gorgeous? Next, use *Find* to go to *Carrara, Italy*. Read the brief narrative about Carrara by selecting *Geography* then look at the quarry image via *Sights and Sounds*. In both locations, rock can be viewed as works of art---in Australia, nature has carved a sculpture from bedrock, and in Carrara, humans take over and shape rocks into museum masterpieces.

8. Which of the three major rock groups are highlighted in both these places? Which types of rocks are found in Katherine Gorge and which are associated with Carrara?

9. Based on the type of rock found in the area, why should you not be surprised that visitors to Katherine Gorge National Park are able to enjoy their vacations on beautiful sandy beaches?

10. What type of material is shipped out of the port city of Marina di Carrara?

11. Who first made Carrara famous for its quarries and who are two famous sculptors who favored the material found there for their masterpieces?

12. Which famous sculpture was carved from Carrara marble, took three years to complete, and was finished in 1504?

Keywords: **igneous, sedimentary, metamorphic, Northern Ireland, Ireland, Katherine Gorge, Carrara**

Exercise 9: Lithosphere and Tectonics
Continental Blue Plate Special: Fettuccine Alfredo Wegener

The dynamics of plate tectonics are complex as they take into account forces that occur both above and below the earth's surface. Some aspects of the tectonic system are clearly evidenced by earthquakes and volcanoes such as the Pacific Ring of Fire. Other aspects of tectonics are less visible on the surface such as subduction, which occurs deep beneath the oceans. In this activity you will study several aspects of plate tectonics and how the affect various regions of the world.

PART I: Tectonic Waters
To begin your examination of plate tectonics, go to *Features* and click on "*Global Themes*." Select "*Physical World*" and then "*Plate Tectonics*" from the list on the right side of the screen. This section will serve as the basis for all of the following questions. Read through the narrative for a review of plate tectonics. Watch the Video Clip at the beginning of the article to view plate tectonics in motion. Now, consider these questions.

1. Who was responsible for developing the theory of continental drift?

2. How was it that a meteorologist came up with the notion of relating atmospheric motion to earth's crustal movements?

3. What type of technological advances were needed to help support continental drift theory?

4. How have studies of the earth's magnetic field assisted in understanding seafloor spreading?

5. What is the Marianas Trench, where is it located, and how was it formed?

6. What are the differences between divergent and convergent plate boundaries? Give some locational examples for each.

PART II: Slide Show: If You Catch My Continental Drift
As you read through the *Global Themes* discussion of plate tectonics, there are two slides that are included that you may click on and view. The first one you encounter is an image of the "*Chilean Andes*" and the second is a photo of the "*Young Himalayas*." Create a map of each mountain chain and then using *Map Styles*, change the map to "*tectonics*." Be sure to read the *Geography* section for both the Andes and Himalayas. The following questions relate to each of these slides, maps and *Geography* information.

7. List all of the tectonic plates that are a part of, or border, South America.

8. Which of these two plates are involved with the formation of the Andes Mountains? Which is being subducted under the other?

9. How have earthquakes helped to shape the Andes Mountains?

10. Where is Cerro Aconcagua and what is its claim to fame?

11. What are some of the valuable minerals found in the Andes Mountains? Why do you think these
    mineral deposits have not been fully exploited, especially in countries that could certainly use the
    economic development that mining would bring?

12. Which two plates are responsible for the formation of the Himalayas? Are these plates converging or
    being subducted?

13. Why are the Himalayas so geologically unstable?

14. Where was the ancient Tethys Sea located and where would you find it today?

15. How do the varying elevations of the Himalayas affect the region's climate and diversity of plants and
    animals?

PART III: Galapagos and Megashear: Really Bad 'Hair Bands' from the '80s?
While you have the *Global Themes* on tectonics handy, click on "*The Galapagos Rift*" and
"*Equatorial Megashear*" that are shown as *Related Topics* in the left hand margin of the screen.
Red through these short narratives and answer these questions.

16. What is the Galapagos Rift zone and what is happening there?

17. Which tectonic plates are involved with the Galapagos Rift and in which directions are they moving?

18. What sort of ocean ecosystem is being created by the Galapagos Rift and what sort of creatures would
    you find there should you take a cruise on the "Alvin?"

19. What is the Equatorial Megashear and where is it found?

20. Which two continents are splitting along the Equatorial Megashear?

21. What is the "Romanche Fracture Zone" and what is its major claim to fame?

Keywords: **tectonic plates, continental drift, Alfred Wegener, Andes, Himalayas, Galapagos,
Megashear**

Physical Geography
Exercise 10: Volcanic and Tectonic Landforms
Return to Cinder

Everyone seems to enjoy the fireworks displays on the Fourth of July. The more awe-inspiring the pyrotechnics the better. Likewise with volcanoes. The ones that draw the most interest are those that spew forth the most hot molten lava high in to the air and then spill over the sides of the mountain. Lava flows then set off forest fires and may even threaten human life and property. Humans see to be drawn to such melodrama and danger, much like slowing down to rubberneck when we see an accident on the highway. Tourists are drawn by the thousands each year to view the remains of Pompeii, the ancient Roman city near present-day Naples, Italy, that was incinerated and buried under tons of volcanic ash and debris. Although difficult to comprehend, some observers were disappointed when in May of 1980, Mount St. Helens erupted, and instead of becoming one great big lava lamp, it only belched ash and thick smoke.

PART I: Watch Where You Step or You May Krakatau
In this activity you will be asked to take a virtual trip around the world visiting different volcanoes and volcanic landscapes. Begin by clicking on *Features* and select *Global Themes*. Under *Global Themes*, go to "*The Physical World*" and click on "*Volcanoes*." Read this interesting discussion on vulcanism and be sure to take a moment to view the *Video Clip* and each of the images along the left side of the screen.

1. How is it that you can compare a volcano to a bottle of champagne?

2. What are the four things that volcanoes emit and give examples for each.

3. What happened in 1902 on the island of Martinique?

4. What happened to Krakatau when it erupted in Indonesia in 1883 and how did this affect other parts of the world?

5. What happened in 1973 on Heimaey that brought both good news and bad news to the Icelandic island? What does Heimaey have in common with Surtsey?

6. How did 1,700 people die in Cameroon in 1986?

7. What is unique about Cotopaxi, a volcano in the Andes Mountains? As you look at this slide, what do you think would happen to the local residents and landscape if this active volcano would suddenly erupt?

113

8. Describe how mudflows can be as dangerous to humans as flowing lava and include an example.

9. What is a strato-volcano and how are they unique as compared to other volcanoes?

10. What is the difference between a cinder cone and a shield volcano and give an example of each?

11. How was the Deccan Plateau of India formed and which North American landform has a similar origin?

12. What are two examples of "dormant" volcanoes that have erupted in the past 20 years?

13. Fortunately for Edinburgh Castle, what sort of volcano serves as its foundation?

14. The Hawaiian Islands are a hot spot for tourists but also for some other reason---what is it?

15. Why don't you find volcanoes associated with the San Andreas Fault?

16. Where is Fuji-san, how did it form according to legend? How does the religious interpretation of this mountain's origin differ from scientific theory?

Keywords: volcanoes, Mount St. Helens, magma, lava, pyroclastics, Krakatau, Iceland, Hawaii, Fuji-San

Exercise 11: Weathering and Mass Wasting
Permafrosting on the Cake:
Physical Processes in Arctic Regions

Much of the northern half of Canada lies within the permafrost zone. The freeze and thaw cycles associated with this climate regime work to create unique arctic landscapes with such features of pingoes, stone rings, and patterned ground. Permafrost also creates difficulties when humans attempt to establish permanent settlements. Careful planning and engineering must be implemented when constructing buildings and infrastructure in order to avoid damage to foundations and underground utilities. With an extremely short growing season, traditional agricultural activities other than animal herding are virtually impossible in permafrost regions. As a result of these and other factors, few large settlements are found north of the zone of continuous permafrost.

In this activity the focus is on how humans cope with permafrost and how permafrost works to create unique landscapes. To accomplish this, we will look at arctic ecoregions in general and then some permafrost environment examples in Canada.

PART I: The World on Ice
For some background information on permafrost and related environmental conditions, go to *Features* and select "*Global Themes*." When you arrive at the Global Themes page, click on "*The Physical World*," then "*Tundra, Polar Deserts, and Ice*." Read through this brief section and be sure to view all four slides that accompany the text.

1. Where do you find the world's two arctic tundra regions? Why don't you find similar ecoregions in the Southern Hemisphere?

2. What are the three contributing factors to arctic tundra environmental conditions?

3. How is it that an arctic tundra can be identified as a "desert?"

4. What happens on the surface during the times that the permafrost begins to partially thaw? How does this help support plant life, birds, and insects as well?

5. What types of plants are found on the Norwegian tundra? Why do you think these plants take the form of small, gnarled, low-lying vegetation?

6. How are the "stone deserts" of the Russian tundra formed?

PART II: Just the Halifax on Permafrost

Canada has many diverse landscapes and one that is most interesting is in the boreal and tundra environments. To take a look at where these conditions occur, create a map of Canada using the *Find* box and change the *Map Style* to depict "*ecoregions*."

7. Using the *Dynamic Sensor*, determine the approximate line of latitude which serves as the boundary between boreal and tundra environments. What was your finding?

8. Now switch your *Map Style* back to "*Comprehensive*." Using the line of latitude that you determined to be the boundary between boreal and tundra ecoregions in the previous question, how many cities do you find north of that line? Which of these cities has the largest population? How do the population sizes of these northern towns compare with those of Edmonton, Winnipeg, and Halifax?

PART III: Sliding Across Canada

While you have a map of Canada on the screen, click on *Sights and Sounds*. Look at the slides titled, "*Canadian Ice Fields*," "*Arctic Moraine on Baffin Island*," "*Inuit of the Canadian Arctic*," "*Mackenzie Mountains*," and "*Ice Floes in Baffin Bay*." With all this valuable knowledge, you are now ready to try your hand at these questions.

9. After viewing the slide, "*Canadian Ice Fields*," would you be interested in taking a tour bus to the ice fields of Jasper National Park? Why or why not?

10. Why do you think hikers only venture into the Auyuittuq National Park Reserve in June and July? If you were talking a hike through this region at that time of the year, what three things you would be sure to include in your backpack?

11. The image of "*Inuit of the Canadian Arctic*" reveals several things about surviving in a permafrost region. Can you spot at least three things in this picture that illustrate how the Inuit have adapted to living in this challenging environment?

12. Why do the Mackenzie Mountains receive such small amounts of precipitation?

13. How does permafrost affect the volume of soil water in the Mackenzie Mountains and how does this work to create rock rings, polygons and patterned ground?

14. It is hard to believe after looking at the "*Ice Floes in Baffin Bay*" slide that William Baffin was able to navigate this region in 1616 and that his record stood for 236 years! How does the northern extent of his journey compare with the line of latitude you determined to be the dividing line between boreal and tundra environments in Part II, Question 1?

Keywords: **permafrost, tundra, arctic, ecoregions, ice fields, Canada, Inuit, Mackenzie Mountains, Baffin Bay**

Physical Geography
Exercise 12: The Cycling of Water on the Continents
My Old Kentucky Sinkhole: Karst Topography

Karst topography is characterized by a variety of surface and underground features but have one thing in common---running water working to erode rock surfaces into such structures as sinkholes, depressions, caves, springs, and swallow holes. Normally, karst landscapes are associated with limestone regions and water flowing in contact with this type of rock dissolves the calcium carbonate of which limestone is composed. Your textbook should provide you with adequate background material on karst landscapes to be prepared for this exercise. In this activity we will learn more about karst features such as sinkholes and caves and take a virtual field trip back in time to Mammoth Cave. So hold on stalactite and enjoy your tour!

PART I: A Cave of Mammoth Proportions
Mammoth Cave National Park is located in western Kentucky and is a popular destination for tourists and scientists alike. To begin your study of this unique karst landscape (above and below the surface), create a map of *Mammoth Cave National Park* using the *Find* command. Take a moment to look over the park's location and surrounding area. Click on *Geography* for a brief description of the region and answer the following.

1. Look at the map of Mammoth Cave and the surrounding region. Find at least four toponyms (place names) that most likely were named after their karst characteristics.

2. Mammoth Cave is famous for some aspect of its configuration--what would that be?

3. What is the difference between a stalactite and a stalagmite? Why are the ones in Mammoth Cave so colorful?

4. What is the "Echo" and what sort of creatures might you find in it?

5. How do you suppose the Green River earned its name?

PART II: Journey to the Center of the Cave
You may have been fortunate enough to have visited Mammoth Cave at some time in your life. But, if you have not, here is your chance to take a virtual tour. And if you have, here is your chance to go back in time and tour the cave through the lens of a 19th century explorer. To begin your historical tour of Mammoth Cave, select *Web Links* from the menu on the left hand side of the screen. Go to the "*National Park Service Guide—Mammoth Cave National Park.*" Once you have the homepage, select "*Mammoth in Depth,*" and then "*A Tour of the Cave---in 1844!*" Follow the directions on how to navigate through the cave and answer the questions below. If you are claustrophic or acrophobic, hold on for all your stalagmite!

6. Who was Dr. Croghan and how is it that he came to write about Mammoth Cave?

7. How did visitors in the 1840s see into the darkness of the cave? Do you think such an illumination system would be permitted in the cave today---why or why not?

8. On the way to the entrance to the cave, guests of Dr. Croghan passed "the ruins of saltpeter furnaces and large mounds of ashes." What was the saltpeter used for? Why did saltpeter processing resume at this location during the Civil War?

9. In the Main Cavern, ruins of hoppers or vats were found. What were miners doing in the cave that required processing in vats?

10. What sorts of graffiti were found in the caves in 1844 and who might have written them?

11. Why were some visitors reminded of the Gothic cathedrals of Europe when they visited Mammoth Cave? How were these much-admired features formed?

12. What did guests usually do when they reached the "Devil's Armchair?"

13. How were the caves used in medical practices and what were they attempting to cure? Do you think this was a good location for such therapies---why or why not?

14. How did visitors get across the "frightful chasm" of the formation known as the "Bottomless Pit?"

15. Why did the tour guide halt the boat trip on the Echo River just three-quarters of a mile into the voyage?

16. How much different do you think a tour of the Mammoth Cave would be today and why?

Keywords: **karst topography, limestone, Mammoth Cave National Park, Kentucky, stalagmites, stalactites**

Physical Geography
Exercise 13: Fluvial Processes and Landforms
Wet and Wild Waterfalls

Some of the most majestic features associated with streams are waterfalls. Breathtaking and beautiful, waterfalls are dramatic elements of many landscapes. In this activity you will travel the world via Encarta Virtual Globe and visit nine waterfalls and explore their unique character and significance. Begin by entering "*waterfall*" in the *Find* box and see what comes up under *Contents*. Select the following waterfall examples from the *Contents* list and answer questions for each.

Sailing in Madagascar
1. How have some waterfalls in Madagascar been used to serve the local populations?

Erawan National Park
2. Waterfalls can be studied as part of the physical as well as the cultural landscape. Why are these falls a good example of how natural features become intimately entwined with local cultural traditions?

Yosemite Falls and Contrasts of Yosemite
3. What distinction does Yosemite Falls have in North America?

4. What geomorphic processes were at work to create these falls?

Icelandic Water Power
5. Iceland's waterfalls could provide this fossil fuel-poor country with an abundant supply of renewable energy, but what might prevent this from happening?

Towering Auyan Tepay and Angel Falls in Canaima National Park
6. What does this Auyan Tepay have in common with Angel Falls (you can click on "Angel Falls" that is highlighted in the caption for Auyan Tepay)?

7. Angel Falls has a special distinction---what is it?

8. Where is the best place to view Angel Falls and why is this so?

Island of Springs
9. Waterfalls in Jamaica are unique not only in their beauty but also in their economic function. How has Jamaica capitalized on waterfalls as part of their tourism economy?

Guyana's Kaieteur Falls
10. What does the word "Guyana" mean?

Click on "King George V Falls" as highlighted in the "Kaieteur Falls" slide. Find King George V Falls on the map. Upon which river do you find King George V Falls and name any other falls you see on that river?

11. What river lies to the east of King George V Falls? What are the names of the falls on that river? Why do you think the river to the east of King George V Falls has more falls?

Concord Falls, Grenada
12. What important function do Grenada's waterfalls serve to the island's population?

Congo River
13. Which river is the only one with higher discharge rates than the Congo?

14. Whereas many of the waterfalls you have looked at so far aid tourism or energy development, the falls on the Congo actually create a problem. What is this problem and how do Congo River users get around this obstacle?

Your Favorite Falls
15. We hate to put you over a barrel, but of the nine previous waterfall sites you have explored, which would you most like to see in person and why?

Keywords: **waterfalls, rivers, tourism, hydropower**

Exercise 14: Landforms and Rock Structure
That Canyon, Steve--It's a Butte!

If you travel to many of America's state and national parks, you know that a visit to one of these areas is like a vacation and a physical geography field trip wrapped into one. Those interested in exploring a wide variety of landform features and rock structures would be well-advised to head for the Rocky Mountain region, in particular, the Colorado Plateau. So pack your bags, your field guides, and some bottled water, and head for the Rockies!

PART I: All Around Four Corners
To begin your working vacation you will need some basic information on the *Colorado Plateau*. Using the *Find* box, create a map of the region. Click on *Geography* for a brief description of the *Colorado Plateau*. You will come back to the *Geography* page for links to other locations later in this activity.

1. Take a moment and look over the map of the Colorado Plateau. What type of stream drainage pattern to you find in this area?

2. Which major rivers have their headwaters forming within the Colorado Plateau?

3. Zoom in on the San Juan River. At this scale, can you explain the number and pattern of intermittent streams?

PART II: Park Yourself in a Canyon
Referring back to the *Geography* page for the *Colorado Plateau* to access these locations (you can click on each park name which is highlighted in the *Geography* text), visit each of the places below and answer the questions for each.

Arches National Park
4. Look at the map of Arches National Park. Why do you suppose a railroad was built from Crescent Junction to the Colorado River?

5. Go to the *Geography* description for Arches National Park. What type of rock has been eroded to form the amazing natural sculptures for which the park is so famous?

Bryce Canyon National Park
6. Click on *Geography* for Bryce Canyon and read the short narrative describing the park. What type of rock has been eroded to form the spires and pinnacles found in there?

7. Go to *Sights and Sounds* and look at the slide for Bryce Canyon. In this case, the park may be misnamed--why?

8. How did Native Americans interpret the Bryce Canyon landscape?

9. There are several unusual toponyms (place names) on the map of Bryce Canyon and surrounding areas. Which one do you think is the most unique and how might it have gotten its name?

Canyonlands National Park

10. Take a look at the map of the Canyonlands National Park and surrounding areas. Based on what you see on the map, why do you think this is one of the least visited parks of those in the Colorado Plateau region?

11. Click on *Geography* for some details about *Canyonlands National Park*. What types of formations do you find in the park?

12. What are petroglyphs?

Capitol Reef National Park

13. Referring both to the map and the brief description of the area you can read by clicking on *Geography*, explain why the park has such an unusual shape—very long and narrow.

14. How did the park get its name?

Grand Canyon National Park

15. Read the *Geography* narrative for *Grand Canyon National Park*. If you were to visit the park, how would you navigate the canyon---what kinds of roads and trails are available?

16. There are four distinct climate zones that have been identified in the Grand Canyon. Explain how such wide variation could occur across the region.

Mesa Verde National Park

17. When you have the map of Mesa Verde National Park on your screen, click on *Geography* and read some details about the region. How did the park area get its name?

18. What is special about Cliff Canyon and Soda Canyon? How were they formed and why were these sites suitable for habitation?

Petrified Forest National Park

19. Don't be afraid—there is nothing scary about this place. Check out the map of the *Petrified Forest* and click on *Geography* for some background information. How did these trees become "petrified?"

20. Take a moment to look at the map. Why do you think this park would have many more visitors than Canyonlands National Park?

Keywords: **landforms, canyons, drainage pattern, national parks, Grand Canyon, Native Americans**

Physical Geography
Exercise 15: The Work of Waves and Wind
Dust Busters: Loess is More

No, we haven't switched from physical geography to the minimalist architecture of Mies van de Rohe. We are talking about the seemingly mundane subject of dust. No, not the dust bunnies under your bed, but rather wind-blown dust that can build to incredible depths in places like the steppes of Ukraine, the plains of the Corn Belt, or the Pampas of Argentina. So what exactly is loess? In this activity you will learn more than you probably ever wanted to know about loess and then some.

PART I: Fine China Soil
For all the dirt on loess, enter "*loess*" in the *Find* box and then click on the glossary entry that appears at the top of *Contents*. Look at the definition for loess and then go back to the *Contents* box. Which country seems to dominate the number of entries for loess? Loess deposits in China are extremely important to that country's agricultural production and partially explains how they are able to produce adequate food supplies for the billion-plus people who live there. Let's take a look at loess in China. Go through each of the entries in the *Contents* box that relate to China and answer the questions below.

1. In northern China, where do you find loess deposits? How has erosion in this region created serious problems for transportation and agriculture?

2. Where do you find loess near the Qin Ling Mountains and what type of agriculture is practiced to its north and south?

3. Where do you find loess in Gansu Province and what types of crops are grown there?

4. Where do you find loess deposits in Henan Province in relation to the mountains? What types of crops are grown in Henan?

5. Why is the Huang He called "China's Sorrow"? The Huang He could also be called China's salvation—why?

6. How did the Huang He earn its name and what might loess have to do with this?

7. What types of crops are grown in Shanxi Province and what measures have been taken to increase production?

123

PART II: Loess Objects

As you went down the *Contents* list, you probably noticed that China is not the only country blessed with highly productive loess deposits. Go back to *Contents* box and click on the images for *Belgium, Kaiserstuhl,* and *Idaho.* Look at each slide, read the captions, and then answer the questions below.

8. Where do you find the most productive farmland in Belgium?

9. If the picture of the Belgium farmland did not have a caption, where might you guess this photo was taken?

10. What type of agriculture is practiced in Kaiserstuhl?

11. What happened to farmland in Kaiserstuhl during the 1970s? What resulted from these changes in crop production?

12. How were Kaiserstuhl farmers able to irrigate their plots without the aid of modern irrigation equipment?

13. How did atmospheric circulation affect terraced farmland in Kaiserstuhl after the hillsides were reconfigured?

14. Where is the Palouse region and what is its claim to fame?

15. Why do you think the Palouse is "one of the nation's least populous?"

16. If you could magically edit the red barn out of the "Palouse Country," which other slide would this landscape remind you of?

Keywords: loess, soil, China, Huang He, viticulture, irrigation, terraces, Palouse

Physical Geography
Exercise 16: Glacier Systems and the Ice Age
The Iceman Misseth: The Driftless Area

Examine any map illustrating the southern extent of the most recent continental glacier and you would notice that the ice sheets stopped at the Ohio and Missouri Rivers. But a feature that is less noticeable on the map is where the glaciers completely bypassed on their trek southward--southwestern Wisconsin, northeastern Iowa, and northwestern Illinois. Glaciologists speculate that the ice sheet broke into two lobes and the so-called "Driftless Area," was simply spared the most recent glacial advance. What happens when an ice sheet bypasses a region? In this activity we will explore the Driftless area and become virtual ice age tourists in three states--Iowa, Illinois, and Wisconsin.

PART I: Iowa: Field of Drifts
If you think Iowa is a perfectly flat state with nothing but endless fields of corn, think again. The northeastern part of the state has a more glacier-free history and the landscape is quite different from the rest of the Iowa. To find out more about Iowa's ice age history and present conditions, create a map of the state by entering "*Iowa*" in the *Find* box. Click on *Geography* and read a brief description of the state's physical characteristics.

1. How did ice-age glaciers thousands of years ago, create favorable conditions for Iowa farmers today?

2. What is the "Young Drift Plains" region?

3. How did the "Dissected Till Plains" get their name?

4. Why do you think the Driftless Area in Iowa has greatest local relief of any region in the state?

5. Decorah, a town in the Driftless Area is nicknamed "Little Switzerland." Obviously someone in Iowa must have had a good sense of humor to come up with that! What would you estimate the elevation of Decorah to be?

6. Dubuque is located farther south of Decorah, on the bluffs of the Mississippi River. One fabled tourist attraction, the Fenelon Place Cable Car Elevator, claims to be the shortest, steepest railroad in the United States. By looking at the map, why do you think this could be true?

PART II: Illinois: Get the Lead Out

Just a short swim across the Mississippi River from northeast Iowa will land you in Illinois' Driftless Area. Go to a map of "*Illinois*" using *Find* tool and click on *Geography*. Read about the physical geography of Illinois. Go back to the map and locate Galena, Illinois, and read about its local conditions by selecting *Geography*. Now you should be ready to answer these questions.

7. What are the three main land regions of Illinois and how are they different?

8. What is the significance of the Central Plains?

9. What type of soil was formed from glacial drift that is so important to agriculture?

10. Other than the Driftless Area of northwestern Illinois, there is another unglaciated landscape in the state---where is it and what is it called?

11. Where is Galena, Illinois located and what was its original name? How do you suppose it earned that ignoble title?

12. What mineral has been mined in Galena (the name of the town should give you a hint) since the 1700s?

PART III: Say Cheese: The Wisconsin Driftless Area

Wisconsin is probably more famous for its Driftless Area in the southwestern part of the state than either Iowa or Illinois. This distinction doesn't come so much from agriculture as it does from tourism. Go to a map of "*Wisconsin*" and read the *Geography* narrative before answering these questions.

13. What are the two main land regions in Wisconsin and how are they different?

14. In which of these two regions would you find the Driftless Area?

15. What type of landscape features would you see if you visited the Wisconsin Driftless area?

16. What is "Fat Man's Misery"---some sort of dinner comprised of nothing but celery sticks?

***Keywords:* Driftless Area, glaciation, glacial till, Iowa, Illinois, Wisconsin, The Dells**

Physical Geography
Exercise 17: Soil Systems
Our Just Deserts: Soil Erosion

If there is one resource we take for granted it would probably be soil. Most of us do not realize that while it can take millions of years for soils to form, soils can be destroyed in a matter of minutes. While erosion and deposition are both part of the natural soil forming process, human intervention can accelerate both with disastrous consequences. Soil erosion is one of the most important environmental issues facing the world today. In this exercise you will look at some of the global issues associated with soil erosion and what can be done to mitigate the problem.

PART I: Soil, Soil, Trouble and Toil
Use the *Feature* button to access "*Global Themes*." When the Global Themes page comes up, click on "*Environmental Issues*" and then over to "*Soil Erosion and Exhaustion*." Read this section and answer the questions below.

1. Over the past few decades, what human activities have caused accelerated rates of soil erosion and exhaustion?

2. Since 1945, what percent of earth's land area has been subjected to soil degradation?

3. How many square miles have been damaged by soil erosion and exhaustion?

4. How many square miles of farmland are abandoned each year due to soil depletion?

5. Why do you think worldwide food production growth rates have slowed since the 1980s? At the same time, what has happened to world population growth?

6. What is the single most common cause of soil loss across the globe? Why does this practice accelerate soil erosion?

7. How does deforestation affect water resources?

8. How have modern agricultural inputs such as fertilizers, pesticides, and tractors damaged soils?

9. In the United States, how has livestock production damaged soils?

10. How much soil does the United States lose each year due to poor farming practices?

11. With all the problems the world experiences with soil erosion and exhaustion, are there any solutions and if so describe these mitigation practices?

PART II: Soil Problems on the Horizon

While you have the page for "*Soil Erosion and Exhaustion*" handy, click on the two slides that accompany the text. You should be able to view images from *Madagascar* and from *Lake Turkana*. When you have looked at both slides, answer these questions.

12. Why has Madagascar experienced such severe deforestation?

13. How do population pressures relate to the rates of deforestation in Madagascar?

14. Where is Lake Turkana? With which river is it associated?

15. What happened to Lake Turkana between 1973 and 1989?

16. What two factors when combined have resulted in increased turbidity in Lake Turkana?

PART III: Desertification: It's Just Not for the Sahara Anymore

Desertification is another critical environmental issue across the world. Read all about it by going back to the *Global Themes* page and selecting "*Environmental Issues*" and then "*Desertification*." Be sure to look at all of the slides that accompany the text and then consider the following questions.

17. What exactly is desertification?

18. We often think of desertification as associated with warm desert environments but this is not always the case. Give an example of how arctic dry environments can also be susceptible to degradation via human interaction.

19. How has irrigation harmed soils and landscapes in arid regions?

20. Where has desertification occurred in the United States?

21. Why is the Aral Sea shrinking?

22. Where and how has livestock raising practices lead to increased desertification?

23. What and where is the Sahel and how have agricultural practices there accelerated desertification?

24. What example is given for reducing desertification in the slide titled "Sound Practices?"

Keywords: **soil erosion, soil exhaustion, agriculture, livestock, desertification, Sahel**

Exercise 18: Systems and Cycles of the Biosphere Global Ecosystems
Act Locally, Think Globally

As we come to the final activity related to your physical geography textbook, we take a different approach to studying the topic at hand. Here is your chance to think globally as you consider how your local geography fits into the big picture. Over the course of these activities, we have touched on many subjects and, it is hoped, raised your interest in environmental issues. Despite the broad scope of these exercises, however, it is unlikely that we have mentioned your hometown or the place where you attend college. In this activity, you will have an opportunity to explore your own world and see how it fits in with the rest of the planet.

Getting to Know Your World
As a class, you should decide which locations you choose to study---each student could look as his or her hometown, or the class could select the town or city where the college or university is located. Whichever you select, just follow the directions for the exercise for that specific location.

1. In which town or city are you studying? In which state or province is it located? Go to a map of your state or province and gather all the information you can about its physical geography and if available, descriptions of your town or city.

2. If you had to choose an environmental issue that you believe most affects your town, what would it be?

3. Go to "*Global Themes*" via the *Features* icon and click on "*Environmental Challenges*." Select one of the categories on the right hand side of the screen that most closely corresponds with the environmental issue you stated in Question 2. Read through your selection (be sure to take advantage of any slides or related topics that you come across) and briefly describe the most important points you encountered in the narrative.

4. As you read through your selected topic from "Environmental Challenges," which part or parts do you think applied to your town or surrounding areas and why?

5. If you were hired by the local government to solve this vexing environmental problem, what solutions would you offer?

Keywords: **environmental challenges, local action, global issues**

Chapter 1: The Study of Economic Geography

One Picture Tells a Thousand Geographies

How often have you spent even a nanosecond of your time studying the cover of a textbook? Okay, so sometimes you don't even want to look at what's inside but just this once, take a few minutes and examine the cover photo on your *Economic Geography* text. What do you see— aside from the title, edition number, and authors' names? If you had to write a story to go along with this image, what would it be? Where do you think the photo was taken and what is its geographic message? Using a little detective work, you will be able to find physical, cultural and economic evidence to support your theory.

Where's Wheeler?

First, focus your attention on narrowing down the locational possibilities for the cover photograph. You should start by looking at a large area and gradually winnow that down to a smaller area. To do this, start with some clues offered by the physical landscape. You may begin by looking at a world map and then if you desire, look at specific regions. Change the *Map Style* first to *Ecoregions* and then *Climate*. This should help you refine your search area. Check the *Legend* to determine what the shadings and symbols mean on the map.

1. Based on what you have seen so far, which world regions have you *eliminated* as possible photo shoot locations?

2. Once you think you have the region that most resembles the physical environment depicted on the textbook cover, *Zoom* in to take a closer look. When you have a map of the general area you are interested in exploring, click on the available references that may help you with your search such as *Geography, Facts and Figures, Sights and Sounds*, or *Web Links*. You could also use the *Find* box to search for specific variables or locations. With this additional information at hand, which region are you investigating as a possible photo location?

3. What can you tell about the people in the photo by cultural clues in the image? Who might the people in the foreground be and what are they doing? Are there any architectural clues that might help you out? If you don't see anything that might match the region you have selected, it is not too late to try another spot.

4. Now let's get down to business—or the economic landscape portrayed in the cover photo. Do you think this is an urbanized country and why?

5. What do you think the level of technology or infrastructure might be in this country? Do you think this is an industrialized economy---why or why not?

6. You have had some time to think about the location of the textbook cover photo, explore some possibilities and narrow your focus to just one country (or if you are really a geography whiz, a single city or region). So we come back to our first question---where do you think the cover photo for your *Economic Geography* textbook was taken and what evidence do you have to support your assertion? You may not have the exact location but hopefully you are reasonably close!

Keywords: **economic geography, ecoregions, climate, culture, economies, urbanization, industrialization, location**

Chapter 2: Global Population Processes and Pressures
Painting the World by Numbers

It may come as no surprise that the table on page 20 of your textbook lists China as the country with the world's largest population. What you may not realize is that the rate of population change for China is less than some other countries on that list. To better understand the world population issue, it is important to compare and contrast demographic characteristics. In this activity, you will be analyzing demographic variables and how they can be used to interpret the differences in global standards of living.

PART I:
We begin our demographic studies by creating a table of statistics to supplement Table 2.1 in your textbook. To do this, you will either need a sheet of graph paper or you could set up a spreadsheet on your computer. Using the list of countries from Table 2.1, create a table using the variables listed below. The data for the table you are crafting can be found by clicking on the *Features* button and selecting *Statistics Center*. When the Statistics Center page appears, click on *Population*, and you will be able to access the data for the following variables for each country. When the map appears, go to the bottom of the screen and click on *Statistic Table* to view the data in tabular form and see each country's world ranking.

Birth Rates
Death Rates
Fertility Rates
Population Density
Population Growth Rates
Percent Rural Population
Food Supply, Daily Dietary Protein
Infant Mortality Rates
Literacy Rates

1. Now that you have all your statistics at hand, how does China compare with other countries—does having a large population automatically reduce the standard of living? Why or why not?

2. In your opinion, which countries on this list face the most serious problems and why?

PART II: The Laws of Averages
Now consider this. All of these statistics are averages and averages can be misleading. For example, imagine you go rabbit hunting. Up jumps a bunny and you shoot with your double-barreled shotgun. Boom! You miss to the right. Boom! You miss to the left. Yet, on average, you hit your target right between the eyes! Or did you? The key to interpreting certain statistics is recognizing that they often fail to show the distribution of phenomena, rather, they tend to show the average of the extremes. Population density is a good example of this. One problem with this statistic is that it is often erroneously used to interpret standards of living rather than what it actually measures—the numbers of persons per square unit of area. The assumption that

is frequently made is that where population densities are high, standards of living are low, and where densities are low, standards of living are high. The logic (twisted as it may be at times) is that in crowded conditions, resources of all types become scarce but where people are spread out over a large area, there is more room to breathe, there are fewer social and economic ills that are associated with crowded urban conditions, therefore, the quality of life goes up. Let's test this notion and see where it often goes terribly wrong---on average!

Go back to the *Statistic Table* and look at the world ranking for *Population Density*.

3.   Which country sits in the number one position---with a bullet?

Click on this country's name and go to a map of this place. Read about this country by clicking on *Facts and Figures, Society*, and *Sights and Sounds*.

4.   How would you describe the standard of living?

5.   Do you think having the highest population density in the world has negatively impacted the quality of life---why or why not?

6.   While there are many countries ranked in the top 20 by population density that have lower standards of living than the United States or Canada, which of those in this upper grouping do you think have similar or higher standards of living than most countries?

7.   Where does the United States rank in terms of population density?

8.   Which five countries rank just higher and which five rank just lower than the U.S.? How do population density and standards of living compare at this level?

9.   Go to a map of the United States (you can simply just click on United States right from the statistical table). Now change the *Map Style* to *Population Density*. Describe the general distribution of U.S. population. Does this accurately reflect the "average" population density and why?

10.   With the same U.S. map, change the *Map Style* again, this time to *Earth at Night*. What does this image tell you about U.S. population distribution and urbanization?

Keywords: **population, demographics, standard of living, China, population density, United States, population distribution, urbanization**

Chapter 3: Global Economic Development
Thumbs Down: A Hitchhiker's Guide to the Globe

If you reside in an industrialized country, you probably take certain things for granted. For example, you might expect to own a car—or even two—and enjoy a high level of mobility. You may also have access to such conveniences as cellular telephones, cable television, computers, and Internet connections. Imagine how difficult it would be to get through your day without electricity, running water, telephones, paved highways, or even for some, your favorite television show. Yet when we look at the 'haves' and the 'have nots' of the world, the majority are 'have nots'. Without money, mobility, education, and information, it is very difficult to break out of the cycle of poverty that grips the vast majority of the world's populations. In this exercise we will look at a variety of economic development indicators, some you might expect, and some that might surprise you, and maybe you will see how lucky you truly are.

Take a Hike
Modern transportation is essential in fostering economic development yet not all countries have adequate infrastructure to allow for the seamless flow of goods, people, and information. Using data from the *Statistics Center*, we will examine how global development has a long way to go and not many good roads to take. To begin, click on *Features* and select *Statistics Center*. Use the information in the *Economic* and *Population* tables to answer the following questions.

Go to the statistical table and under the *Economic* category, look at the world rankings for motor vehicles per 1000, telephone mainlines per 1000, and television receivers per 1000.

1. List the top five and bottom five for each category in the space below.

2. Referring back to your lists from Question 1, what do the countries that ranked in the top five categories have in common? What about those that ranked last for each of the three groups?

3. According to the data for motor vehicles, which world region would probably be the worst to travel in if you were going to hitchhike your way across the countryside? Why do you think this part of the globe has so few motor vehicles?

Now switch your statistical table over to the "Population Category." You will need the world rankings for the following categories. List the top and bottom five for each:

Electricity Consumption Per Capita

Fax Machines per 1000 People

134

Internet Hosts Per 10,000 People

Mobile Telephones

Personal Computers in Use

4. Which countries consistently show up in the top five rankings for the categories listed above? Based on what you know about these places, what do you think is the relationship between communication and economic development?

5. And now for something completely different. Look at the statistics for *"Cinemas, attendance per capita."* What are the top five countries and why do you think the number one country has so many theater goers, a much greater percentage of the general public than there are movie fans in the United States or Canada?

6. Does the country that ranks number one in *"personal computers in use"* also rank first in *Internet connections*? Why do you think the top five countries for computer use do not lead the world in Internet hosts and what is the regional pattern of the top Internet users? Why do you think people in this part of the world are spending much of their time on line (and on mobile phones!)?

***Keywords*: economic development, transportation, infrastructure, motor vehicles, communications, computers, Internet, cinemas**

Chapter 4: The Interdependent Global Economy
Banking on Technology

Think of yourself as a citizen of a developing country with an annual per capita income of $1000. How would you afford a college education for yourself or your children? With that much income, would a bank loan you money to buy a tractor or more land for your farm? Most likely, the bank's loan officer, after reviewing your application, would show you to the door. Just such lack of investment capital is one of the major reasons why some regions tend to lag behind others and people have such a difficult time overcoming the vicious cycle of poverty. One approach taken by countries outside the industrialized realm to deal with the lack of investment capital and low levels of technology is to pool their resources as a group. One such organization is the Group of 15 (G-15). In this activity you will learn more about how countries cooperate to share technology and economic development strategies through structures such as the G-15.

G-15...G-24...BINGO!
To access information about the G-15, enter "Group of 15" in the *Find* box. You will notice that another "G" appears under contents along with the G-15—the "Group of 24." Click on both, read the descriptions for each, and answer the following questions.

1. How would you describe the regional membership characteristics of the G-15? Are there any North American or European members---why or why not?

2. What is the Non-Aligned Movement (NAM) and why do you suppose it has such a large membership list?

3. What is meant by coordinating "South-South" and "North-South" links as one of the main objectives of the G-15?

4. One of the major accomplishments of the G-15 has been the establishment of the South Investment Trade and Technology Data Exchange Program (SITTDEX). What is the purpose of this program?

To judge how successful SITTDEX has been in the adoption and diffusion of computerized technology among the G-15 members, go to the *Statistics Center* and see how each member ranks in terms of Internet host and computer users.

5. Which G-15 member countries rank the highest for each statistical category? What do you think are some of the barriers to widespread usage of computer networks in G-15 countries?

6. What is—or should we say---what are the Groups of 24 and how are they different by membership and purpose?

7. If you were sitting on the board of directors of the G-15, what would you propose as the three most important economic and technological issues facing the member nations and how these problems might best be solved? Naturally, there are no easy answers but you should be able to come up with some possibilities.

Keywords: **technology, finance, economic development, G-15, G-24, NAM, SITTDEX, computers**

Chapter 5: Principles of Spatial Interaction
Geography is Very Spatial To Me!

When Taaffe, Morrill, and Gould developed their now-famous network model, they based their ideas on transportation development and change in colonial West Africa. Thus, this model is best applied to situations where coastal settlement occurred first followed by interior explorations. Take a moment to review the Taaffe, *et al.* network model in your textbook (Figure 5.9).

1. Could you apply the model developed for West Africa to an examination of transportation systems development in the United States today? Why or why not?

States in the United States may be at different levels of economic development and integration in the transportation system. We will examine the four states of Alaska, Minnesota, Tennessee, and New Jersey with respect to their degree of transportation system integration. First, create a map of 'Alaska' using the *Find* option. Zoom in on the *Seward Peninsula* near Nome (you could use the *Zoom* in function or the *Find* command).

2. Beginning with Nome and working eastward, write down the names of all the towns you find along the way until you reach Golovin or even Elim.

3. How would you describe the transportation network in the southern half of the Seward Peninsula?

4. Would it conform to the scattered port stage of the Taaffe, Morrill, and Gould model? Why or why not?

Click on the largest city in the Seward Peninsular region. It also has the highest route density within the local area. Now go to *Web Links* and select *Alaska Internet Travel Guide.*

5. Discounting RV travel, what type of transportation is promoted at the very outset when you click on *'Transportation'* at this Web Site?

6. Why does this mode of transportation seem favored over others?

7. What factors continue to inhibit highway network development in this region and all of Alaska?

Now take a trip to Duluth, MN using the *Find* tool and typing in 'Duluth'. Read *Geography* to get a better understanding of how the twin port cities of Duluth, Minnesota and Superior, Wisconsin increased in size and economic significance by being connected to interior resources.

8. What types of products are shipped through Duluth and Superior?

9. How do these products differ (if at all) from those shipped from the ports on the Seward Peninsula?

10. Does the transportation network in the Duluth-Superior area conform to Taaffe, Morrill and Gould's model stage of lines of penetration? Why or why not?

Tennessee's transportation network is used as an example of the network interconnection phase (i.e., feeder lines extend from interior cities thus facilitating the expansion of their *umlands*). Create a map of '*Tennessee*' using the *Find* command and, using the sidebar labeled '*Geography*', determine the three largest cities in the state. Click on each of these three cities in turn to learn more about their economic profiles.

11. Which of the three has the most network connections and why?

12. Predict which Tennessee city is poised for the greatest growth thanks to its position within the network and elaborate your reasoning.

To see a state that would display high priority linkages, we can use New Jersey. Access a road map of New Jersey through the *Web Links* for that state. Select *State of New Jersey* and then click on *Transportation*. Through this maze of federal, state and local roads you'll find several high priority highways including: I-95, I-195, I-295, I-80, and the New Jersey Turnpike.

13. Several major port cities are linked through this network. Can you name them?

The New Jersey highway system provides crucial network connections between Pennsylvania, New York, Delaware and, of course, New Jersey itself. These arteries keep the 'Boswash' megalopolis going.

***Keywords*: Taaffe, Morrill and Gould network model, transportation, Alaska, Duluth, Tennessee New Jersey, umland**

Chapter Six: The Role of Transportation in Economic Geography

The Root of the Canal System: Erie, Isn't It?

The Erie Canal is the hallmark of Edward Taaffe's "trans-Appalachian era" of transportation development in the United States. The Canal was completed in 1825 and had a dramatic effect on the flow of goods and people from east to west. So grab your mule and let's take a trip on the Erie Canal. Locate the canal by using the *Find* command. To tour the canal and the surrounding regions, select *Web Links* and click on *Erie Canal History*. Next, scroll down to link to the *New York State Canal System* official homepage (http://www.canals.state.ny.us/canals/) and click on *Canal History* to read all about the development of the canal and its subsequent economic and social impacts.

1. What advantages did the canal have over turnpike routes like the National Road?

2. Where were the Northwest Territories and why would anyone want to go there?

3. What is a granary?

4. Why did New York governor DeWitt Clinton envision New York City as the "granary of the world?"

Look at Figure 7.5 in your textbook.

5. How does New York's import hinterland relate to its position on the canal system?

6. How is the population distribution of New York state influenced by the canal even to this day?

7. What event in 1959 led to the decline of the Erie Canal's importance?

8. What types of economic activities are most common along the Erie Canal today?

The Erie Canal had a tremendous influence over the fortunes of Buffalo, New York. It became a break-of-bulk (i.e., transshipment) point based on its site and situation along the canal system. To get a feel for life in 19th century Buffalo and western New York, while still on the web links associated with the map of the Erie Canal (*Canal History*), select the *Web Link* entitled *Canal-Related Links* and click on the one marked *The Buffalo History Works* (mistakenly labeled *Erie Canal homepage*). Once there, click once again on *The Erie Canal*. Examine the sites within that link entitled *The Buffalo Harbor* and *The Infected District* (http://bhw.buffnet/erie-canal) and answer the following questions.

9. How did the natural environment, finances, and politics help establish Buffalo's position on the canal?

10. How did the waterfront area get the nickname, "The Infected District?"

11. How did railroad network improvements change the role of the Erie Canal in Buffalo?

12. Why isn't there a National Football League team called the "Black Rock Bills?"

***Keywords*: Edward Taaffe, trans-Appalachian, Erie Canal, New York City, Buffalo, New York**

Wheeler, et al., *Economic Geography*, 3e
Chapter Seven: The City as an Economic Node
Houston, We Have a Problem: Invasion of the Multiple Nuclei

Of the three classical models of urban land use discussed in the textbook, the "multiple nuclei model" of Chauncy Harris and Edward Ullman may be the easiest to apply to contemporary urban landscapes.

1. What are the four reasons why the land use patterns associated with Ullman and Harris's multiple nuclei model supposedly emerge? (HINT: See p. 146 in the textbook for the answer)

Let's apply these reasons to the city of Houston, Texas. Create a map of Houston using the *Find* command. Then click on *Geography* to learn about the city's morphology (form). In particular, focus on the discussion of zoning ordinances.

2. What is the nature of zoning in Houston and why is it so?

Print out the map of Houston that is displayed within Encarta. Now, click on *Web Links* and then open *City of Houston—Internet Connection*. Click on *Virtual Tour* and use this map to help you identify Houston's land use patterns. Check also the *City of Houston* web site for additional information about the city. Now click on *Houston Resources* and select *Greater Houston Partnership*. There are several useful links available through *Overview*. Excellent information about Houston is also found on *Map of Key Locations* and *History*. Print out the map available at this web site for future reference. Click on the sidebar called *Sights and Sounds* to take a look at two aspects of the city. All of these maps, images and written information will provide the necessary background to apply multiple nuclei theory to a real-life city.

3. Referring to the textbook's discussion of classical urban spatial structure especially Figure 7.7, try to identify the land use "districts" of Houston.

4. What type of residential neighborhoods would you expect to find near a museum?

5. How about near a shipyard?

6. How about near a heavy industrial area?

Using one or more of the maps you have printed off, circle and label the areas (zones) you can delimit within Houston.

7. Does the emerging spatial pattern resemble the model of the multiple nuclei? Why or why not?

8. Would the concentric zone or sector theory models be a better fit for Houston? Why or why not?

9. How do you think the lack of zoning ordinances throughout most of Houston's history has affected its land use pattern?

Unless you have spent time in Houston, you may not get all the land use patterns exactly right but you should be in the right ballpark—or at least near the Astrodome!

Keywords: classic models of spatial structure, Chauncey Harris, Edward Ullman, multiple nuclei, land use, Houston, Texas

Chapter Eight: The Location of Tertiary Activities

Here's the Rub: The Dutch Have to Modify Central Place Spacing

It isn't often that a theory developed under ideal circumstances and assumptions is applied verbatim in the real world. But, that is exactly what happened in the Netherlands when the Dutch wished to settle the land area that they had reclaimed from the sea. In 1932, a very long earthen berm (i.e., embankment) was created to enclose a former inlet of the North Sea.

1. What is the body of water once called the Ziederzee by the Dutch now called ? (HINT: To find out, click on *Find* and bring up a map of The Netherlands. Go to the sidebar marked *Land and Climate* and click on it to find the answer)

2. What do the Dutch call the large bodies of land that they have been able to reclaim from the North Sea?

3. How much land has been reclaimed from the sea since the 1930s?

Emmeloord is located where the first parcel was drained and reclaimed. Click on *Find* for that city and bring up a larger scale map of this area to the east of Amsterdam. Can you see the long earthwork that was built to hold back the angry North Sea?

To get an inkling of what it was like to drain such a mucky, brackish (part saline) area click on *Sights and Sounds*. Now click on the slide entitled *Gathering Rushes*...

4. How does that caption to that slide read?

5. How does the action of farmers in the 1950s bypassing shopping at the smaller, nearby agricultural service centers established by the Dutch government in favor of further but larger centers both within and outside the region confirm Rushton's "improved behavioral postulate" of central place theory?

Spacing among the remaining settlements on the oldest reclaimed land is still relatively close. To find out how close, move the mouse to the pull-down menu at the top of *Encarta* called *Tools* and click on *Measuring Rule*. Use the mouse to locate the town of Urk as the origin and Emmeloord as the destination.

6. What is the straight-line distance that intervenes?

7. Now measure and record the straight-line distance between Emmeloord and Lemmer. What is it?

Compare those figures with the spacing between settlements on one of the more recent land reclamations. Focus on the largest center in this part of Flevoland called Lelystad. Measure the distance between Zeewolde and Lelystad using the *Measuring Rule*.

8. What is that distance?

9. What is the distance between Lelystad and Dronten ?

10. Approximately how much further apart on average are central places in this new polder than they are in the original?

The Netherlands learned from its mistake of slavishly conforming to the distance of the classical theory. Lelystad must be upgraded to the next level of the central place hierarchy and must serve as a regional center for the three reclaimed areas to which it is centrally located. Theory must be modified to account for dynamic changes in behavior within central place systems.

Keywords: **Netherlands, polders, central place theory, Lelystad**

Chapter Nine: The Changing Economic Geography of the Restructured Metropolis

A Tale of Three Cities

Far from being the bedroom communities built in the image of Levittown, New York, contemporary suburbs are just as likely to be a place where you commute to work. In fact, the most common commuting pattern today is from suburb to suburb rather than the traditional pattern of suburb-to-central city. Table 9-1 in the textbook (pp. 181-82) reinforces the losses that central cities have suffered to their suburbs in three employment categories—total employment, employment in manufacturing , and employment in retail trade between the years 1976 and 1994.

The city with the highest overall job loss to suburban employment was New Orleans. Use the *Find* command to focus on the city. Then click on *Geography* sidebar and read some background material on the city.

1. When did population and subsequently jobs begin their exodus from New Orleans?

Denver lost a whopping 34 percent of its manufacturing employment to its surrounding suburbs between 1976 and 1994. Create a map of Denver and zoom in for a close-up look. Click on the communities of Aurora, Englewood, Lakewood, Littleton, and Wheat Ridge, using the *Geography* sidebar to find out more about these Denver suburbs.

2. What types of products are manufactured in the Denver area?

3. Does the nature of what is being produced foster the migration of manufacturing away from Denver?

4. Which Denver suburb is the largest in population?

5. Which suburb has the most diversified economy?

6. Which suburb is poised to create the stiffest competition for future jobs and why?

Of all the cities listed on Table 9.1 of the textbook (pp. 181-182), Baltimore has lost the greatest percent of its retail trade to its suburbs (22 percent). What makes Baltimore so vulnerable to suburban retail centers? In order to answer this question, create a map of Baltimore using the *Find* command. Take a tour of the city via the following URL (http://travel.roughguides.com/). This web address will take you to the home page of the always interesting travel guides entitled *Rough Guides*. There is a pull-down screen at the top of the home page called *Featured Cities*. Open it and scroll down until you find *Baltimore*. Read the interesting information about Baltimore contained on this overview and then click on the area marked *Districts*. When that screen comes up, click on *Downtown Baltimore* and then click on the *Go* button underneath. Read the associated text and then try to answer the following questions.

7. How did the fire of 1904 change the downtown?

8. Why did it fail to recover?

9. How does the author of this page describe the city's downtown shopping district?

10. What recent projects in downtown Baltimore have taken place in lieu of retail development?

11. Is this the wave of the future for the city or are they just fishing around for a solution to avoid becoming another living urban museum?

Two cities that fared the best in the urban-suburban competition for retail employment were New York and San Francisco. New York City lost only ten percent of its retail trade jobs to its suburbs and San Francisco only five percent.

12. What advantages do these high profile coastal cities have over, say, Baltimore? Material in Encarta can be an invaluable resource in formulating your response.

Keywords: **Baltimore, Denver, New Orleans, New York City, San Francisco, suburbanization**

Chapter 10: Manufacturing: Where Plants Locate and Why

Run for the Border: NAFTA

This activity is designed to assess the impact of the maquiladora (twin plant) operations along the US-Mexican border. Specifically, this activity focuses on the border towns of El Paso, TX and Cuidad Juarez, Mexico just across the international border. Besides lower wages in Mexico, the proximity to markets and transportation in the U.S. and beyond has lured investment in Mexico by American-owned businesses.

In many ways, the border region has been an industrial target region for years as manufacturers have been attracted to the available and inexpensive labor market in Mexico but also by the weak environmental regulations and enforcement. It is hoped that the environmental policies of NAFTA will result in cleaner industries along the border but so far progress has been slow.

Create a map of Cuidad Juarez using the *Find* command and then click on *Web Links*. Go to *Cleanup Along US-Mexican Border Remains a Dream* and read about some of the serious water quality and health hazards affecting Juarez and other US-Mexican border towns (e.g., Tijuana, Nuevo Laredo). Now click on the link entitled *Silva Reservoir* contained within the longer reading.

1. What role is NAFTA playing in the fight to clean up residential and industrial waste?

Now, go to the *Find* command and produce a map of *Tijuana*. Click on *Web Sites* on the sidebar and select *Tijuana River Pollution and Maquiladoras*.

2. What is perceived as the main impact of Tijuana River pollution?

3. Why is California concerned about pollution in Mexico?

4. How has the pollution affected the health of children on both sides of the border?

Now travel (virtually at least) south from Juarez to the Mexican city of Chihuahua (i.e., use *Find* type in '*Chihuahua*' the city and capital of Chihuahua state). Go to the *Web Links* sidebar and then click on *Chihuahua Industrial Parks*.

5. Which American-based companies have located manufacturing plants in the state of Chihuahua?

6. Is Chihuahua's growth related to NAFTA?

Now cross the border (virtually, that is) to El Paso, Texas. Click on the map symbol for *El Paso* and go to *Web Links*. Select *El Paso* and find out more about the city. On the web page, click on the *Maquila* link.

7. What is the purpose of "bonding" a maquiladora?

8. Why would the Mexican government want to enforce such a policy?

9. What location factors attracted manufacturers to this region?

10. What is the basic and non-basic employment impact of maquiladoras in Juarez?

11. Are there any negative impacts of maquiladoras mentioned on this site?

12. Why do you think that is?

***Keywords:* NAFTA, maquiladora, manufacturing location, Juarez, El Paso, Tijuana, environment, pollution**

Chapter 11: Manufacturing: Regional Patterns and Problems

Diamonds in the Rough: Facets of Industrial Development

Do you think you might know which country ranks first in percent of gross domestic product (GDP) earned from industrial output? You can check your intuition by creating a world map and selecting *Statistics Center* under *Features*. Select *Choose Statistic* and click on *Economy* and then *Gross Domestic Product, Industry (no date)*.

1. Which country is number one?

Were you surprised to see that it was Angola and that the United Arab Emirates (UAE) were in second place? Scan down the list of countries and find the highest ranking European country for which data have been included. Let's find out why these countries are ranked so high.

2. Which European country (of those reporting the statistic) leads all others in Europe?

Go to a map of Angola using the *Find* command. Check out the *Facts and Figures* sidebar for some background information on the country.

3. What are the major industries in Angola?

4. If 75 percent of the population earn their living from agriculture as is so symptomatic of third world countries, how do you explain the high GDP per capita from industry? What sectors and activities constitute 'industry'?

Click on the *Society* side bar and scroll down to read the *Infrastructure* section.

5. How does Angola's economy compare to other African nations?

6. What appears to be the biggest problem facing Angola's progress toward development?

7. Why do you think a large middle class has yet to emerge in Angola?

Now, go to the *Web Links* sidebar and examine three web sites—*ABC News Country Profile: Angola, Encarta Online—Angola,* and *Angola—A Country Study*. In all three, pay particular attention to the discussion of the country's economy.

8. What product accounts for about 90 percent of Angola's exports?

9. Why doesn't Angola mine the diamonds that are so plentiful in the northeastern part of the country?

10. Where is Cabinda and what is an exclave exactly? (HINT: You may have to go to the *Find* command and type in *Cabinda*)

11. If Cabinda province were contiguous to the rest of Angola, would the Democratic Republic of the Congo (formerly Zaire) be landlocked without access to the sea? What about the Republic of Congo?

12. Which foreign country has made the greatest investment in Angola's manufacturing industries and what is the primary export market?

13. What are some of the positive and negative aspects to investing in Angola?

Now, let's compare Angola to the Czech Republic, Europe's leader in percentage of their GDP generated by industry of the European countries reporting such information. Create a map of the Czech Republic and then click on the *Facts and Figures* sidebar.

14. Do you see any similarities between Angola and the Czech Republic in terms of exports?

15. How do the two countries compare in basic demographics?

16. Which country has the higher per capita GDP?

17. How do the population growth rates and infant mortality rates compare?

18. Which of these two countries has, in your opinion, the highest potential for further industrial growth?

19. In which sectors will this growth occur and why?

***Keywords*: manufacturing, Angola, petroleum, diamonds, GDP, Czech Republic**

Wheeler, et al., *Economic Geography*, 3e

Chapter 12: The Economic Geography of Energy

Hot Under the Crust: Geothermal Energy in Iceland

The more we can exploit alternative energy sources such as solar, wind and geothermal power, the less dependence there will be on fossil fuels to supply the electric power we need. Figure 12.17 in the textbook shows potential geothermal sources of power in the United States.

1. Which appear to be the three leading states in geothermal potential in the continental United States?

2. About what percentage of the energy in the United States is generated by geothermal sources?

The country of Iceland is a great deal more dependent upon geothermal sources than we are in the United States. Create a map of *Iceland* and select *Land and Climate*. Click on the first slide you see entitled *Icelandic Water Power*.

3. Why do you think that Iceland has failed fully to take advantage of its hydroelectric power potential?

Read about the country's *Economy* by clicking the *Society* sidebar. Next, click on the *Sights and Sounds* sidebar. As you "tour" the island (virtually, of course), pay special attention to the ways that volcanoes and volcanic landscapes have shaped settlement patterns and economic activities.

4. How are hot springs utilized by Icelandic farmers to increase food production and grow exotic crops such as bananas in such a decidedly non-tropical location?

5. What sorts of natural hazards are associated with geothermal regions?

6. How have these natural hazards both helped and hindered development in Iceland?

7. What is possibly the only world's capital to be heated entirely by geothermal steam?

8. Which city in the United States cited in your textbook had a similar system as early as the 1890s?

9. Do you think that geothermal steam is still used for that city's heating needs? Why or why not?

Find out what alternative energy source is being touted by an Idaho utility company by going directly to its homepage. The URL is as follows: (http://www.idahopower.com/) When at the homepage of the *Idaho Power Company*, click on the sidebar icon for *Energy Resources* and then click on photovoltaic (solar) within the text. The page devoted exclusively to *Solar Energy* will come up. After reading the material, answer the following questions.

10. What is a photovoltaic system and why might such an energy source appeal to some Idahoans?

11. Why, in your opinion, would a public utility become involved in this type of technology?

12. What advantages would solar power have over using geothermal power in Idaho?

13. Do you think that Icelanders will be converting to solar power in the near future? Why or why not?

Keywords: **geothermal power, energy alternatives, tectonic activity, volcanoes, Iceland, Idaho**

Chapter 13: The Spatial Organization of Agriculture

"Real" Agriculture: Implosion or Explosion?

Figure 13.22 in Wheeler, *et al.,* (p. 326) is a rendition of a map that first appeared in John Fraser Hart's article entitled "A Map of the Agricultural Implosion"[1] in which he tried to define operationally what was meant by real agriculture. He didn't want to include the dabblers, the hobby farmers and others who earned most of their income off of the farm. Lacking any conceptual guidelines, Hart chose to map the percentage of farms in a county possessing gross incomes of more than $10,000 from the sale of farm products. Only if that percentage exceeded fifty percent was the county deemed worthy of inclusion in his definition of a "real" agricultural county. As an can be seen by a perusal of Figure 13.22, real farms are rather limited in areal extent. Five areas stand out most markedly—1) the Corn Belt states of Iowa, Illinois, Indiana and nearby adjacent states; 2) the Central Valley of California, an irrigated paradise of specialty produce, orchard crops and ranching; 3) the truck gardening areas of New Jersey (the Garden State), southeastern Pennsylvania and the Delmarva Peninsula; 4) the Atlantic Coastal Plain region of the Carolinas known for cotton, tobacco and hog production; and 5) the lower Mississippi River valley famous for its soybeans and cotton. Is it still true that agribusiness (i.e., Hart's 'real' agriculture) is more concentrated than agriculture as a whole?

Let's look at the most up-to-date statistics about agriculture on the web obtained from the 1997 Census of Agriculture. Using the *Find* command, type in '*United States*' and then examine the associated *Web links*. Click on the link to the *US Census Bureau*. Then click on *Subjects A-Z* and click on *Agriculture*. Scroll down the options until you come to *NASS* (National Agricultural Statistics Service). Within *NASS,* click on *Census of Agriculture* (Starting with the 1997 Census of Agriculture, the NASS of the USDA has been responsible for its production rather than Bureau of the Census of the Department of Commerce). Then click on *1997 Agricultural Atlas of the United States.* Now, click on *Go to Maps Index Page* and then scroll down until you see the category entitled *Farms by Value of Sales.* Examine (i.e., print off if you are able) three dot distribution maps for the three highest categories of Farm Sales—*Farms with Sales of $100,000 to $249,999: 1997; Farms with Sales of $250,000 to $499,999: 1997; and, Farms with Sales of $500,000 or More: 1997.* Because of the diminishing number of farms, the number of farms represented by each dot decreases as the monetary value of farm sales increase.

1. Where are the areas of high concentration of farms with large sales?

2. How well do these areas match the ones that Hart came up with using different statistics for the 1964 Census of Agriculture?

2. Assuming success can be equated to farm sales, what differences do you detect between the pattern of modestly successful farms (i.e., those in the $100,000-$249,999 category) and the extremely successful farms (i.e., those in the $500,000 or more category)?

---

[1] John Fraser Hart, "A Map of the Agricultural Implosion," *Proceedings of the Association of American Geographers*, Vol. 2 (1970), p. 69.

A better comparison with Hart's county-level data is obtained using two other statistics that the NASS has recorded by county: 1) *Percent of Farms with Sales of $500,000 or More, 1997*; and 2) *Value of Sales of Farms with Sales of $100,000 or More as Percent of Total Value of Sales: 1997.* Only 2.4 percent of farms nationwide had sales of $500,000 or more in 1997, but some counties had well over nine percent of farms that fell into the category (the darkest black value on the map).

4. Are these counties, for the most part, in the region defined by Hart as practicing real agriculture in 1964?

5. Are there some areas of the country that have been added or in which the value of farming has intensified over the last thirty years?

Interestingly, it appears that the Corn Belt of the agricultural Middle West, while a very important agricultural region, is still the domain of the large family-owned farm corporation. The area has fewer farms with sales over $500,000 than, say, the specialized agricultural cornucopia in the Central and Imperial Valleys of California which appear to be dominated by really large growers (i.e., agribusinesses).

What crop and/or livestock combinations are represented in the following five areas that seem to have experienced a considerable amount of agricultural intensification in the past thirty years?

6. The floodplain of the Red River (of the North) that forms the boundary between North Dakota and Minnesota?

7. The Imperial Valley of extreme southeastern California?

8. The Texas panhandle region?

9. Areas of eastern Washington and Oregon extending into the Snake River plain of southern Idaho?

10. Central and Southern Florida?

Keywords: **Agricultural Implosion, Agribusiness, United States, Central Valley (California), Corn Belt**

Chapter 14 Contemporary American Agriculture
How Now Dairy Cow?

It may be time to loosen the traditional "dairy belt". The major dairy states measured by number of cows or pounds of milk produced are no longer necessarily located in the historical dairy belt as illustrated by Figure 13.17 of your textbook (p. 319). The pattern as shown does not include some major dairy states outside the northeastern core region. When you think of dairy production, which state comes to mind? Why Wisconsin of course—America's Dairyland. It says so right on their license plates. And indeed, Wisconsin did rank number one in the number of milk cows in the most recent United States Department of Agriculture census. But the Great Lakes states do not rise to the top as the crème de la crème when total milk production or amount of milk produced per cow are considered. We can use Encarta Virtual Globe '99 to learn more about the structure of the American dairy industry until the cows come home.

Create a map of *Wisconsin* using the *Find* command, click on *Web Links* and select *State of Wisconsin Information*. From that web page go to *Wisconsin State Agencies* and select *Agriculture, Trade and Consumer Protection Department*. Then, select *Wisconsin Agricultural Statistics Publications* (*WASS*) and scroll down to *Dairy* reports and click on *Annual Milk Production*, a report produced on February 18, 1999. You will be able to milk this page for gallons of information!

1. What are the top six dairy producing states in the United States in 1998? (HINT: Read the textual material preceding the tables—don't rely on the first table by itself because it is incomplete)

   1.
   2.
   3.
   4.
   5.
   6.

2. Of the states shown in the table, which state ranks first in the average production of milk per cow?

3. Of the states shown in the table, which state ranks first in the millions of pounds of milk produced?

4. Given that in the 1990s, the Wisconsin cow produced 15,442 pounds of milk per year, and the average person drinks two eight-ounce glasses of milk per day, how long would it take one person to consume the milk produced in one year from just one Wisconsin cow?

Keywords: dairy belt, fluid milk, Wisconsin, dairy cows

Chapter 1: Regions and Themes

Despite Its Many Faults, California is Everything It's Cracked Up to Be

Californians may be a bit sensitive to the fact that 90 percent of all earthquakes in the continental United States occur in their state. There are indeed many faults in California from the best known San Andreas that runs from Los Angeles to San Francisco to the most newly discovered parallel fault lines in the desert some 55 miles southeast of Los Angeles. But, California is so much more than the occasional earthquake. In a drive of 85 miles one can go from the highest point in the continental United States—Mt. Whitney at over 14,000 feet—to the lowest point in the desert country of Death Valley more than 200 feet below sea level.

It is small wonder that the motion picture industry grew to one of major proportion in southern California at the beginning of the twentieth century. The low-technology film of the era required the bright sunshine of southern California's Mediterranean climate for proper exposure. Also within a sixty-mile radius of Hollywood one could find seashore, desert, mountain ranges or fertile valleys in which to shoot a variety of movie genres "on location".

If you examine closely the map depicted in Figure 1-2 (p. 9) of Birdsall, *et al.*, you'll see that few other states in The United States or provinces in Canada display such a degree of regional overlap as does the state of California. Within the state are portions of regions 13 (Empty Interior), 14 (Southwest Border Area), 15 (California) and 16 (North Pacific Coast) as defined by the authors. Associated with each of these regions is a particular theme (or themes) that the authors of the textbook wish to emphasize (see Table 1-1 p. 8).

1. For each of the four regions, list the topics that the authors wish to emphasize.

Part I. Photographic Images of California.

Now, let's turn to the material in Encarta '99 to sample some of California's vast diversity as a state of overlapping regions and environments. Let's first take a close look at the slide set associated with the state of California. Type in *'California'* in the space next to the *Find* command. Then click on *Sights and Sounds*. You will see a series of eight color slides of California. Some of these sights will confirm the stereotypes you may already hold of this state about which everyone seems to have a strong opinion.

2. Which two slides represent, in your opinion, the epitome (or essence) of California? Defend your choices.

3. Which two slides most surprised you? That is, which two slides went the furthest to disabuse you of stereotypes of California you may have previously held? Defend your choices.

Now let's examine the eight slides individually starting with the one in the upper left hand corner and proceeding to the one in the lower right hand corner.

4. Slide 1: How much rainfall does Death Valley normally receive?

5. How can plants and animals survive in a harsh environment where temperatures have been known to soar to 57 degrees Celsius (135 degrees Fahrenheit)?

6. How many unique species of plants and animals are found only in Death Valley?

7. Slide 2: Which features (cultural traits) give this photograph away as representing Chinatown in San Francisco?

8. Slide 3: What does the Golden Gate Bridge connect to on its northern end?

9. Slide 5: Which physical features form the eastern boundary of Los Angeles?

10. When was Los Angeles founded?

11. Slide 6: Who is Anaheim's most famous citizen? How long has he "lived" there?

12. Slide 7: What is the largest alpine lake in North America?

13. How deep is that lake?

14. How much snow do the surrounding Sierra Nevada Mountains receive?

15. How might this snow pack impact the decisions of a farmer in the Central Valley of California several hundreds of miles away?

16. Slide 8: What explorer witnessed surfing in both Tahiti and the Hawaiian island of Oahu as early as 1777?

17. Why was surfing banned in the 1820s and only revived again in this century?

Part II. Examining Web Links for even more detailed images of the Golden State.

Which website would be a better link for information about a state than a site maintained by the state itself? Click onto *Web Sites* and then proceed to the *State of California Home Page*. Now click on *History and Culture: People of California* and then click on *Symbols and Images*. Now click on *California Photographs* and you will see twelve photographs meant to represent twelve distinctive subregions of the state. Click on the subregion called *Desert*. This subregion is closest to the textbook's region 13 (Empty Interior).

18. Which state park and national park are located within this subregion?

Likewise, the subregion called *Shasta-Cascade* may be closest to textbook's region 16 (North Pacific Coast).

19. What is the name of the state park depicted in this region?

The region called 'California' in the textbook might best be represented by that agricultural cornucopia we know as the *Central Valley*. There are several shots of Fresno.

20. Why do you think that the production of table grapes or the processing of another agricultural product is not shown in any of the slides?

Finally, the ethnic diversity of the Southwest Border Area (region 14) might be illustrated by some of the photos in *Orange County*. Garden Grove, for example, has one of the largest concentrations of Vietnamese population outside the country of Vietnam.

21. And yet what is illustrated in the photograph for that town?

22. To what extent does an impressive edifice such as the Phillip Johnson-designed Crystal Cathedral reinforce our image of a region?

23. If the State of California had wished to emphasize the ethnic diversity of the southern California area, what images might have been more appropriate?

Keywords: **California, regions, desert, Sierras, Central Valley, Orange County**

Birdsall, et al., *Regional Landscapes of The United States and Canada*, 5e

Chapter 4 Megalopolis

Urban Behemoths: Wave of the Future or Relicts of the Past?

French geographer Jean Gottman first introduced the term "Megalopolis" to the urban literature in the early 1960s with the publication of his book by the same name. At the time he saw one megalopolis coalescing from Boston, Massachusetts to Washington, DC. Subsequently, other authors and urban growth patterns have extended the southern boundary of Megalopolis to the Norfolk-Virginia Beach-Hampton Roads area of the Virginia Chesapeake. Let's turn to Encarta '99 to see how prevalent the term coined almost 40 years ago has stood the test of time. Using the *Find* command, type in *Megalopolis*. Notice that the term has entries in both *Places* and *Content*. Because it is there, let's focus for a second on the town of Megalopolis.

1. What is another name for this town?

2. What is the name of the mountain range to the northeast of Megalopolis?

3. Do you think that Megalopolis suffered any damage from the devastating earthquake of August 1999? Why or why not?

4. About how far it is from Megalopolis to the capital city of the country? (Hint: you may have to change the scale of maps by sliding the cursor along the Map Scale Bar).

5. What are some features that the town, despite its name, lacks in order to be considered a true megalopolis?

Now let's focus on the meaning of the term 'Megalopolis'. Return to the menu by clicking on *Find* and look up the definition of *Megalopolis* in the glossary. In the sidebar beside the formal definition is a site within Encarta Virtual Globe '99 where you can turn for even more information on this topic. Click on *Urbanization and Cities* and scroll down to the section entitled *The Process of Urbanization*.

6. What does the term Megalopolis mean in Greek?

7. Name two other megalopolises besides the one that extends from Boston to Washington and beyond.

Using the *Find* command, return to click on 'megalopolis' and examine the photographs and descriptions associated with slides that include the term 'megalopolis' somewhere in the accompanying textual material and then answer the following questions.

8. How extensive is the southern California megalopolis?

Curiously, one of the cities mentioned along with the term megalopolis is one that you have probably never heard of. Go to the slide of Tlalnepantla.

9. Of what megalopolis is it a part?

Now, using the *Find* command and *Population Density* as a *Map Style*, let's examine three potential megalopolises—one in the Western Hemisphere and two in the Eastern. Bring up a map of *India* to begin with. With the *Map Style* set on *Population Density*, use the cursor to encompass a rectangular area from Calcutta on the southeast corner to Lahore in Pakistan on the northwest. Follow the India-Bangladesh border northward toward the Himalayas and thence northwesterly toward Lahore.

10. Does it appear that there is almost continuous urban development between the two cities listed above?

11. What are some of the cities that are listed within this corridor?

Now, bring up a *Population Density* map of *China* and answer the following questions.

12. Is there a Pacific Rim Chinese megalopolis emerging between Bejing at the north and New Kowloon to the south?

13. Is there any extensive region along the coastline where the population density wouldn't constitute designation as a megalopolis?

Finally, let's turn our attention to South America and more specifically the area along the Atlantic coastline of Brazil, Uruguay, and Argentina from Rio de Janeiro to Buenos Aires. Bring up a *Population Density* map of *Brazil* and then use the cursor to make a rectangle that would include the above mentioned cities.

14. At this geographic scale of analysis, would you call the coastal region between the two above-mentioned cities a megalopolis? Why or why not?

15. If you answered negatively, can you tell from other *Map Types* what might cause a barrier to continuous or almost continuous urban settlement at relatively high densities?

Keywords: **Megalopolis, conurbation, Boston-Washington corridor, coalescence**

Chapter 5 North America's Manufacturing Core
Gritty Cities and Rust Belt or Nexus for a Brighter Future?

At the turn of the 20th century, over 75 % of the nation's manufacturing took place in a roughly parallelogram-shaped area bounded by Milwaukee, Wisconsin on the northwestern corner, Boston, Massachusetts on the northeastern, Washington, DC on the southeastern, and St. Louis, Missouri on the southwestern. But, even then there were ominous signs of change in the wind. As early as the 1880s, many textile mills began moving out of New England and the Middle Atlantic states for the lower wages and sunnier climes of the Piedmont South. The leakage of industry outside of the traditional industrial manufacturing belt soon turned into a hemorrhage, and by the early 1950s, only about one-half of the nation's manufacturing took place in the "manufacturing belt" (approximately one-seventh of the nation's land area).

It is impossible to assess the impact of this industrial flight and the often painful conversion to a post-industrial society that all cities in the manufacturing belt have experienced since the 1950s, so let's concentrate on two—Pittsburgh, Pennsylvania and Cleveland, Ohio. These cities are not randomly selected. Pittsburgh, the archetypal steel town, was once hailed as "hell with the lid off" by one author who viewed the shower of sparks given off into the night air by the molten steel being smelted and formed in the many iron and steel blast furnaces of the era. This was a time when steel mills worked round the clock to fill orders for I-beams for skyscrapers, steel rails for the nation's extensive network of railroads, ingots of steel to be shaped into wire, cans, refrigerators, and a thousand other consumer products. Steel was to the turn of the 20th century, what advances in computers and information technologies are to the turn of the 21st century. Likewise, Cleveland, OH was a steel-making center. Cleveland was advantageously located because of its site along America's fourth seacoast—The Great Lakes. The Great Lakes and the relative cheapness of water-borne transportation over long distances, made it possible for iron ore from northern Minnesota to be shipped great distances to Midwestern processing centers like Cleveland without greatly adding to the overall cost of production. By the end of the 1970s, plant closures and structural changes in manufacturing left both cities foundering. But, being larger metropolises—the Cleveland-Akron-Lorain, OH Consolidated Metropolitan Statistical Area (CMSA) is number 13 in population within the United States and the Pittsburgh-Beaver Valley, PA CMSA is number 19—both these urban areas pulled on other sectors of the economy to make the transition to a post-industrial economy. Pittsburgh no longer has an active steel mill within its city limits, but it is world renowned for its work in medical research and robotics among others. Cleveland was once derided as the "mistake by the lake", a city that dumped so many toxic pollutants into the Cuyahoga River and nearby Lake Erie that the river actually caught fire! Now with its Rock and Roll Hall of Fame, new inner-city stadium for the Cleveland Indians (Jacobs Field) and a revitalized downtown and waterfront, it is often touted for its high quality of life (i.e., urban livability).

Let's use Encarta's Virtual Globe '99 to get a better idea of what we might expect to find in these two cities that are shedding their "rusty" image. Use the *Find* command and create a map of *Pittsburgh*. Then click on the *Web Links* sidebar and open a web site called *City of Pittsburgh*. Once there, click on the area that will take you to the *Mayor's Office*. We are most interested in the topic of *Economic Development* so click on it.

1. What appears to be the new theme (logo) for Pittsburgh's business community?

Within the *Economic Development* homepage there is a link to *Why Businesses are Choosing Pittsburgh*. Click on that and answer the following questions:

2. What are the five major factors cited for why businesses are choosing Pittsburgh?

3. What *Fortune* 500 firms are located in Pittsburgh? How do you explain the metropolitan area ranking 19th in population, but 7th in the number of *Fortune* 500 corporations?

Now, return to the portion of the *Economic Development* page that discusses *Industry Cluster Information* and answer the following questions.

4. How many Pittsburgh-area firms are involved with environmental technologies?

5. What specific Pittsburgh-based universities are cited as crucial to biomedical research?

6. How many biomedical technology venture firms have been founded since 1996 in Pittsburgh?

7. What biomedical pioneer is associated with Pittsburgh? What did he develop in 1954?

8. What place in the United States does Pittsburgh hold when it comes to employment in high-tech computer software development?

9. What major telecommunications company has located its International Operations Center in Pittsburgh?

10. What are the reasons cited why manufacturing has once again become a growth area in the Pittsburgh economy?

Now, using the *Find* command, make a map of *Cleveland, OH*. Now using *Web Links*, open the link simply entitled *Cleveland, Ohio*.

11. What appears to be the new logo phrase of Cleveland? (HINT: It is not "The Fictional Home of the *Drew Carey Show*")

There are several links at the top of the *Cleveland, Ohio* homepage that we might wish to explore. Let's start with *A Great Place to Work*. Many of the subheadings will take you to tables of (dry) statistics. One that is a little more interesting is called *Proximity to Market*.

12. Just how centrally located is Cleveland, OH according to this information?

Now click on *Major Development Projects* and answer the following question.

13. List a few of the more important projects begun in the 1990-94 time period.
One of the projects was *Jacobs Field*, one of several inner city major league baseball stadiums built with the intimacy of older ball parks and which serve as an anchor in the downtown area. One can access information about the park by clicking on *Cleveland Indians* within the *Development Projects* homepage. Once on the *Cleveland Indians* homepage, click on the sidebar called *Jacobs Field* and answer the following question:

14. How many fans have passed through the gates of the new field since it was opened in 1994?

Another major project begun in the 1990-94 period and finished in 1995 is Cleveland's Rock and Roll Hall of Fame. Access this homepage by clicking on the site under *Development Projects* or returning to the *Cleveland, Ohio* homepage and clicking on *Great Place to Rock and Roll*. Scroll down and read *Frequently Asked Questions (FAQs)* about the Rock and Roll Hall of Fame. Now answer the following questions:

15. Why was the Hall of Fame built in Cleveland (i.e., what is Cleveland's claim to rock and roll fame)?

16. What world-famous architect designed the building?

Now let's go on a *Virtual Tour of the Facility* and answer the following question:

17. Who is the Exhibition Hall named after? Why?

Keywords: **Cleveland, Pittsburgh, post-industrial, manufacturing belt,** *Fortune* **500**

Chapter 11 The Agricultural Core

Agricultural Surpluses: Mutiny over the Bounty?

An interesting statistic about American agriculture is that it is so efficient and so commercialized that the two to three percent of the gainfully employed population who are farmers produce, on average, 120 percent of the country's needs for food. This means, of course, that foodstuffs represent one of the major exports the United States has to offer. One export item in demand overseas is soybeans. Two of the leading soybean- producing states are Iowa and Illinois.

Let's see what we can find out about soybeans (especially the export market). Use the *Find* command to make a map of *Illinois*. Now click on *Web Sites* and go to the official *State of Illinois* web site. There is a *Search* function on that site's home page. Type in *'soybeans'* and more than 200 articles, legislative acts, and other data are presented. Let's examine two of them. The second set of ten articles contains one entitled *On the Farm Questions*. Scroll down until there is a question about soybeans. On the way to soybeans you will see that Illinois farmers are even dabbling in emus and ostriches on a commercial basis.

1. Where does Illinois rank in soybean production?

2. What products are made from the soybean? Does this surprise you?

In the third group of 10 reports is one called *Illinois Department of Agriculture Ag Exports*. Examine the data and answer the following questions:

3. What three countries account for the largest share of Illinois' soybean exports?

4. What three countries account for the largest share of Illinois' soybean meal exports?

5. What three countries account for the largest share of Illinois' soybean oil exports?

Interestingly, the State of Iowa does not maintain a web site that has as much information about agriculture as does the State of Illinois. So, for comparative statistics, we will have to turn to the federal government. Since the 1997 Census of Agriculture, the responsibility for conducting the collection of agricultural data has been turned over to the Department of Agriculture. Earlier Censuses of Agriculture were conducted by the U.S. Bureau of the Census, a division of the Department of Commerce. To access data on agricultural output and the like, log onto the Department of Agriculture's web site and more specifically to the homepage of the National Agricultural Statistics Service (NASS) at http://www.usda.gov/nass/ Once on the homepage of this agency, click on *Search* and type in *'soybeans'*. One of the sites that will come up is a series of reports (actually Excel spreadsheets) entitled "*Crops by State*." When you click on that series of reports, you are faced with a "code" that you must crack. If you are looking for soybean production by state for the years 1988-1997 you must click on *sb198897.csv*. Good for you. You've broken the code—sb means soybeans; and 198897 means the 1988-1997 period. Now, it is probably safe to open the files on your computer although you do have the option to save them to disk first. It is doubtful that there would be computer viruses contained within official documents of the federal government but one can never be too careful! When you've opened the Excel spreadsheet, write down the heading of each column of data and scroll down to the data for Illinois (IL) and Iowa (IA) and answer the following questions:

6. In what year between 1988 and 1997 was the market price (as measured in $/bushel) highest and what was the market price in Illinois and Iowa that year?

7. In what year between 1988 and 1997 was the physical yield of soybeans (as measured either in acres harvested or production in thousands of bushels) the greatest in the two states?

8. In what year between 1988 and 1997 was the market price of soybeans the highest? In that year, how would you describe the yield of the crop?

9. In there an identifiable relationship between quantity of physical output and market price that might be detected? If so, how would you describe it?

10. Give a quick perusal to the data for other soybean growing states in the United States. In general, how do the yields in Iowa and Illinois compare to other states?

11. Do a mathematical calculation to determine what proportion of the total value of the crop of soybeans was accounted for by just the states of Iowa and Illinois in 1997. (HINT: You will have to know the total value of production for the United States which is at the end of the spread sheet)

Soybeans are grown because they are a legume (i.e., they are capable of fixing nitrogen from the soil into their root system thus reducing the amount of artificial fertilizer that needs to be applied). They grow well in both humid continental, warm summer climates (e.g., the Corn Belt of the Middle West), and humid subtropical climates (e.g., along with cotton in the lower Mississippi River valley). They have a fairly short growing season so that they can be planted in late spring if there is an unusually damp or cold early start to the typical season. They can be converted into hundreds of products including vegetable oil for human consumption and meal that can be fed to livestock. They are considered an important source of protein in the diets of many Pacific Rim cultures eaten in the form of tofu. They are used as meat extenders in the United States, though they are not usually consumed as directly as they are in the Pacific Rim countries. And, finally, soybeans have never been subject to any federal acreage restrictions.

Let's view graphically the information collected by the Department of Agriculture about soybeans. Return to the homepage of the *National Agriculture Statistics Service (NASS)* at the URL given above. Now click on the Sidebar marked *Agricultural Graphics*. When that page appears, click on *Field Crops* and scroll down to *Soybeans*. You will see that there are several graphs and maps that have been prepared dealing with soybeans. Study them carefully and answer the following questions:

12. How would you typify the period from 1979 to 1990 in terms of soybean production? What about the period from 1991-99?

13. Where were the greatest declines in acreage devoted to soybeans between 1998 and 1999 taking place?

14. What state had the great yield of soybeans in 1999 as measured by the number of bushels per acre?

15. Given that the total production of soybeans in 1999 topped 2.78 billion bushels, how many bushels might have been harvested if every state were as productive as the most productive one? (HINT: In 1999 how many acres of soybeans were harvested?)

16. What quarter of the year are soybean stocks typically at their highest level? At their lowest?

Now, return to the *NASS homepage* and look under *Frequently Asked Questions*. Click on the site entitled *Slide Show, Quick Facts from the Census of Agriculture*. Watch the slide show and answer the following questions about the graphic displayed:

17. What has been the trend in the number of farms in the United States from 1967 to 1997?

18. What is the modal category (i.e., the one accounting for the greatest percentage of farms) of farm acreage in the United States? What percentage of farms are larger than 2,000 acres?

19. Between 1992 and 1997 what can be said about the average age of farmers in the United States?

20. Interestingly, the value of crops produced in 1997 ($98 billion) was almost equal to the amount of livestock and poultry sold that year. What percentage of crop sales was accounted for by soybeans? What is the only crop that contributed a larger percentage of sales?

Keywords: **Middle West, soybeans, Corn Belt, National Agricultural Statistics Service (NASS), US Department of Agriculture**

Chapter 16 The North Pacific Coast

This Isn't Tan, It's Rust: Raining on the Emerald City's Parade?

Cities in the Pacific Northwest consistently rank very high on studies of the perceived quality of life in the United States. The highest on this list is usually "the Emerald City" of Seattle. It is called the "Emerald City" largely because of the intense green of the Douglas firs and other vegetation that grows so abundantly in the region. Interestingly, the city of Seattle receives no more rainfall annually than many areas in the eastern United States. But, much of the rain Seattle does receive is in the form of low intensity drizzle. Strong frontal activity involving thunderstorms is much more frequent east of the Mississippi River. Hence a favorite Seattle gag T-shirt line: This isn't tan, it's rust. The Marine West Coast climate of the Pacific Northwest is mainly influenced by air masses forming and moving east from the cold North Pacific Ocean. Of course, cold air masses cannot hold as much moisture as warmer air masses so the probability of a lot of rain in any single event is relatively low.

What are the consequences of a lot of cloud and overcast days in Seattle? One thing is that the city leads all others in the number of pairs of sunglasses purchased per capita. When the sun does shine on those rare, festive occasions, the citizens of the Emerald City have forgotten where they put their last pair of sunglasses and hence find the need to buy another pair. Seattle also leads all cities in the proportion of cars ordered with sun (or moon) roofs. If it is sunny, residents of Seattle want to make the most of it. In that regard they are no different than Swedes who vacation in the (often-sunny) Balearic Isles of the Mediterranean.

Let's take a look at the Emerald City. Click onto the sidebar marked *Sights and Sounds* and be prepared to do both—see and listen that is. Make sure that your speakers are turned on to hear a selection of the music that, in the late 1980s and early 1990s made Seattle famous—grunge rock. The clip is from one of the less well-known tunes from one of the best-known bands—Soundgarden. Now, singer Chris Cornell of Soundgarden, has recently released a solo album with quite a different sound than the grunge that made him famous.

Surprisingly, only one slide is shown of Seattle and it isn't even the stock photo of the Space Needle, a feature left over from the 1962 World's Fair. One does see the Public Market sign signaling the entrance to the Pike Street Market, a city-subsidized market for the sale of vegetables, fruits, seafood, and a variety of local products that has become quite a tourist attraction. And yet, unlike the former Fulton Fish Market in New York, Quincy Markets in Boston, or the Inner Harbor region of Baltimore, the Pike Street Market in Seattle has retained more of the original atmosphere of the market. That is, it has not been converted completely to festival (i.e., themed) retailing as James Rouse and his associates have done with the three landmark sites outside of Seattle.

Encarta Virtual Globe '99 is a product of Microsoft and it fairly well known that Microsoft's headquarters is in the Seattle area and that Bill Gates' fabulously expensive and electronically sophisticated dream home is located on Mercer Island. Can you find it? Mercer Island that is. Return to the map of Seattle and answer the following questions:

1. What about the relative site and situation of Mercer Island would make it a very desirable place to live within the greater Seattle metropolitan area?

2. Is Mercer Island more accessible than the equally beautiful Bainbridge Island or any number of other emerald islands that dot the Puget Sound lowland?

3. Microsoft's headquarters has a Redmond, Washington mailing address. Can you find Redmond (HINT: you may need to use the *Zoom* function—the *magnifying glass* + command—to get a better look at the relative location of Redmond).

Is grunge rock really dead in Seattle now as some have claimed? Go to the *Web Links* and click on one entitled *Seattle Sidewalk*, a site which contains the latest in entertainment. Scroll down the sidebar until you come to a heading marked *Find a ...* and then click on *Music event*. When that screen comes up, let's perform a little experiment. For the *region or neighborhood* category, let's keep it at the default category of *All*. For the *music event category*, use your mouse on the pull-down menu and click on *Alternative*. For the *special option* category let's keep it at *All* and for the *time frame* category let's use the mouse to scroll all the way down to *Next Three Months*.

4. How many shows are listed for this *Alternative* category over the next three months?

5. What are the venues? Do the places sound like big amphitheaters or relatively small clubs?

6. Now, just for fun, switch the musical category to *Latin, salsa* and keeping all the other categories the same, how many shows of that musical genre are scheduled for the next three months. Does this surprise you? Why or why not?

Seattle. It's not just for grunge anymore.

Keywords: **Seattle, grunge rock, alternative music, Pike Street Market, Mercer Island, Microsoft, Emerald City**

Chapter 1: Latin American Issues and Chapter 2: Physical Environments of Latin America

Where *Not* to Be in Raincoat Sales: High and Dry in the Atacama Desert

The Atacama Desert of northern Chile is one of the driest places on the planet as a result of a combination of sub-tropical high pressure and cold water upwelling associated with the Peru Current. Some locations in the 600-mile long desert region such as Calama have never recorded any rainfall at all! But how do humans cope with the nearly non-existent precipitation and why would they want to live there anyway? Let's start to answer this question by locating our study area. Use the *Find* command to create a map of the *Atacama Desert*. Click on '*Antofagasta*' and select *Geography* to read about economic characteristics of the area and why anyone would choose to reside in a region that receives 0.5 inches of precipitation per year. Mining is a major activity in the Atacama. Just north of Calama is the mining town of Chuquicamata. Use *Find* to locate *Calama*. Move your cursor and click on *Chuquicamata*. Then select *Sights & Sounds* to see why Chile is one of the world's major mineral producers.

1. What do you suppose might be the environmental and social costs for this type of resource exploitation?

Now, move back to the main map of the Atacama Desert. Then, select *Sights & Sounds* and take a peek at the *Atacama Desert Rings*. Could this location be featured on the next episode of the X-Files as proof-positive of alien contact? This bizarre looking landscape doesn't offer too much in the way of human habitation, but from it you can learn much about the biogeography of the Atacama plateau.

2. Would you expect the Atacama to experience such low temperatures allowing for these types of formations?

Now select *Web Links* and click on *Plants of the Atacama Desert*. The URL has recently changed to (http://www.sacha.org). Scroll down the sidebar until you come to a description (and beautiful pictures) of some exceptional flowers that bloom in the coastal sands of the Atacama. These are the *Coastal Bromeliaceae*. Skim the article and answer the following questions.

3. Why do ecologists describe *Lomas Formations* as fog oases?

4. What is *phytogeography* and how can it be used to interpret environments and climate in the Atacama and elsewhere?

Compare the survival strategies developed by desert vegetation to those used by humans as described in the *Water Innovations in the Atacama Desert* web page.

5. Would it be safe to say that in this case for those adapting to the harsh realities of the Atacama, it's a good thing to be in a fog?

Keywords: Atacama Desert, Chile, Mining, Climate, Biogeography

Chapter 3: Aboriginal and Colonial Geography of Latin America
Trail of Years: Historical Linkages Along The Inca Highway

Much like the Romans centuries before, the Inca developed an extensive empire in Andean America. Military power, food production and distribution were made possible by, and supported through, an amazing transportation network high in the Andes Mountains. How did they do it? Through the *mita* labor system, males were conscripted to work a portion of the year in the military or on such public works projects as road building. The map in your textbook on page 53 (Figure 3.2) provides insight as to the extent of the highway network. Imagine the difficulty of carving and maintaining a roadway in such high altitudes, subject to adverse weather conditions, using 15th and 16th century technology. Then there was always the risk of a fourteen-llama pileup on the Cuzco throughway!

Let's start our journey on the Inca highway at Cuzco (also spelled Cusco), centrally located along the transportation network. Use *Find* to create a map of *Cuzco* and the surrounding area. Click on *Geography* for the surrounding area (*Cusco* as it lies in the *Cordillera Oriental*) get a feel for the history and physical setting of the city.

1.   How does the name of the city itself reflect its significance to the Inca Empire?

Now take a virtual tour of the city by selecting *Sights & Sounds*.

2.   What cultural clues can you gain from this series of slides and narratives that illustrate the confluence of indigenous and colonial cultures?

3.   Listen to the Bolivian panpipe music. How have the Quechua- and Aymara-speaking peoples retained their language and music traditions when indigenous peoples elsewhere have struggled to maintain their cultural identities?

4.   How do you think the diffusion of Incan culture, as well as that of the Spanish, was influenced by the transportation network?

5. Compare the map on page 86 (Figure 3.7) with that of the Inca highway map on page 53. How did colonials adapt the Inca transportation network to suit their own needs?

Along the trail is one of the most famous Incan ruins, Machu Picchu (also spelled Machupicchu). Today, tourists usually launch their trek to the ancient ruins from the city of Cuzco. To get a better understanding of the site and its relationship to the Inca trail, go to *Find* and type in *'Machu Picchu'*. Here you will *find Web Links*. Select the site entitled "The *Inca Trail* and Machu Picchu" identified within the text.

6. Would you walk that many miles for a llama?

***Keywords:* Inca, Machu Picchu, Cuzco, transportation, Andes Mountains, diffusion**

Chapter 4: Agriculture

Coca Puffs: Cracking Down on the Cocaine Trade

Without a doubt the most profitable agricultural products for some Latin American countries are also highly illegal in most places. The illicit drug trade between Columbia and the United States has been well documented and has been a source of friction between the two countries. Americans have encouraged farmers to grow alternative crops such as cut flowers, another high-value yet much more acceptable export product.

While Colombia is a major cocaine refiner, it is not the leading producer of coca. This distinction rests with Peru. Peru is the world's leading producer of coca leaves. Using the *Find* command, create a map of *Peru*. Click on *Society* and go to the section on *Infrastructure*. For additional information on Peru, go to *Web Links* and review the page entitled "*Peru: Consular Information Sheet*".

1. From what you have read, what factors contribute to the drug trade in Peru and do you think this will continue as a chronic problem?

Next, let's compare conditions in Peru to those in Colombia. Locate *Colombia, South America* by using the *Find* option. For an overview of the country's economy, click on *Society*, and then *Economy* for background statistics and descriptions of agriculture. Colombia produces a wide variety of crops thanks to varied topography, soils, and climate. Consuming coca leaves, in a variety of forms, in an ancient tradition. Yet coca plays a central role in the contemporary illegal drug trade in the U.S. and elsewhere. Return to the map of Colombia and continue to explore this issue using the *Web Links*. Select, "*Colombia--A Country Study*," then scan down the list and click on "*Drugs and Society*." Read through the narrative keeping in mind how the geography of Colombia has played a key role in the illegal drug trade.

2. What crop became popular to produce for urbanites by the 1930s and what were the results of its expansion?

3. How does drug crop cultivation affect food production?

174

4. How did the cocaine trade develop in Colombia?

5. What were the early connections between Colombia, Cuba, Chile and the United States?

6. What were the societal consequences of the drug trade in Colombia?

7. What is *basuco* and how has it lead to serious problems in Colombia?

8. Based on what you now know, do you think there are any solutions to the drug problems within Colombia?

9. How would these solutions be similar or different from solving the global drug trade dilemma?

Keywords: **Peru, Colombia, coca, cocaine, illegal drug trade, marijuana,** *basuco*

Chapter 5 Population: Growth, Distribution, and Migration

Why You Have to Go to New York City to Get Your Go-Go Boots Shined: Brazilians on the Move

The mobility of Brazilians has changed the demographics within the country as well as in regions well beyond its political boundaries.   The overwhelming trend in Brazil over the past century has been for rural migrants to seek their fortunes in urban areas.   This has lead to spectacular urban population growth rates far exceeding their rates of natural increase.   Brazil has also been a destination for immigrants as well as a source of immigrants.   So what exactly do Brazilian migration and go-go boots have in common? Let's go to Encarta and find out.

First, create a map of Brazil using the *Find* command. Select *Society* and read the sections on *Demographics* and *Infrastructure*.

1.   What are the major ethnic groups in Brazil?

2.   Which Asian nationality is well represented in Brazil and where do they live?

3.   Where would you find the largest African Brazilian population concentrations?

4.   How has migration within Brazil changed the lives of indigenous peoples such as the Yanomami?

Next, let's switch to *Web Links* and click on "*Brazzil*." When the homepage for this on-line Brazilian culture magazine comes up, select *Search* and enter the word "*migration*." You should get a table of at least 15 articles with a focus on migration. Read the following articles selected from the table and answer the corresponding questions.

"I See Gold in Your Future"

5.   Where is Serra Pelada (you can use the *Find* command to locate it on the map) and why would 80,000 migrants want to go there?

6.   What are the conditions in the mining camps in Serra Pelada?

7.   Have men and women had equal opportunities to make their fortunes?

"American Workers, Exiles, and Emigrants in Brazil"

8. What is *jeitinho brasileiro* and why might it be more important for Americans to grasp than Portuguese?

9. Why do most Americans immigrate to Brazil?

10. When did this pattern of migration begin and where was the major destination?

11. What are some of the cultural adjustments to be made by American expatriates?

"Favelas Commemorate 100 Years"

12. What is a favela and why would few Brazilians enthusiastically celebrate this special anniversary?

13. How did the favelas get their name and who is "credited" with establishing them?

14. Who lives in favelas and how does their standard of living compare to other urban dwellers?

15. What is Rochina's claim to fame?

16. Why are the favelas of Rio de Janeiro situated on hills?

17. What did urban planners have in mind for favelas over the years?

18. What types of social class rivalries have developed between favelas?

"News From Brazil---Brazilian Emigration March 96,"

19. Prior to the 1960s, where were the source regions for immigrants to Brazil?

20. What happened in the 1960s to change that pattern?

21. Why do you think Brazilians are leaving the country in the 1990s?

22. Where are Brazilians choosing to live in the United States?

23. How has chain migration reinforced the pattern of Brazilian settlement in the United States?

24. What are the occupational and educational differences between male and female Brazilian immigrants?

25. What types of acculturation problems do Brazilian expatriates encounter in their adopted homelands?

26. What is life like for an illegal Brazilian immigrant in the U.S.?

27. Why are some Brazilians moving to Japan?

28. How are their experiences different or similar to those who chose the U.S.?

29. Why would Brazilians seek political asylum in Great Britain?

30. Why does Florida lead in the numbers of Brazilians behind bars?

31. Who are the *Brazucas* and where is "Little Brazil?"

And after reading this final article, you now know why New York City, shoe shines, go-go boots and *Brazucas* go together!

Keywords: **Brazil, migration, immigration, emigration, favelas**

Chapter 6: The Latin American City

Show Some Spine! Land Use in the Latin American City

As shown in Figure 6-6 (p. 177) of the textbook, the land use structure of Latin American cities is quite different from that typically found in cities of North America or Europe. In North America the poorest people often live in inner city neighborhoods fairly close to the downtown or central business district (CBD). By the same token, rich people often live at the periphery of the city in large houses on estate-sized lots. This leads to a fundamental paradox: poor people living on very expensive centrally located urban land and wealthy people living on relatively inexpensive land at the periphery. This paradox has led one urban scholar to suggest that in North America, accessibility to central places must be an inferior good while quantity of land itself must be a superior good. What are inferior and superior goods? Think of steak and potatoes. Generally speaking, as the income of a person rises his or her propensity to eat steak increases (especially some of the tenderest, juiciest cuts). Likewise, as a person's income rises, his or her propensity to eat potatoes, a cheap source of starch and calories in the diet, tends to decrease. We say that steak is a superior good, whose consumption increases with increasing income, whereas potatoes are an inferior good whose consumption decreases with increasing income. Can you think of other examples of superior and inferior goods?

Latin American cities (and cities throughout third world countries for that matter) do not manifest this paradox. In Latin America, the wealthy often live in central locations in residential districts close to shops and services of all kinds. Alternatively, the poor are forced to live on marginal land at the periphery of the city often in slum-like conditions. These areas are known by a variety of terms in Latin America—favelas in Brazil, poblaciones caiampas ("mushroom settlements") in Chile and, more generally, squatter slums elsewhere. Using the *Find* command, type in *'Mexico City'*. Now click on the sidebar entitled *Sights and Sounds* and read the caption on the slide designated *Urban Hardships*. These marginal areas are often, surprisingly, slums of hope rather than of despair. As bad as conditions are in these places that often lack basic services such as sewage treatment and piped potable water supplies, they are often better than the conditions in the rural countryside from which many of these recent immigrants arrive.

The Griffin-Ford model accounts for changes in these squatter slums. Eventually, through the dint of effort of their residents, these shantytowns are upgraded—the tarpaper shack gets siding, a new roof, an inside toilet and piped water. This is called *in-situ accretion* in the model. Whole neighborhoods slowly become less tenuous and more permanent in their characteristics until a point is reached in which it is difficult to tell that the area was ever a peripheral slum area. In the Latin American city, accessibility to the central area is a superior good. Rich residential areas such as the Polonco district in Mexico City are close to the Zona Rosa, a major shopping district in downtown Mexico City and not far from the green and leafy Chapultepec Park. Using the *Find* command, select *Mexico City*. Then bring up a map of the city's street layout and general configuration using the *zooming tool* to focus on small areas in greater detail.

1. Do you see Chapultepec Park? How close is it to the main square (the Zócolo)? Hint: One of the buildings on the Zócolo is the National Palace (the Palacio National).

Like many of the Law of the Indies towns, Mexico City's true heart is the main square (the Zócolo) not far from the center of the ancient Aztec capital Tenóchtitlan from which modern Mexico City grew.

If the Zócolo is the cultural and governmental center of Mexico City, where is *the affluent spine* of the Griffin-Ford model? Most likely it would be a broad avenue that leads away from the Zócolo.

2. Is the avenue on which the main entrance to Chapultepec Park is located part of the affluent spine?

3. How about the diagonal avenue trending northeast of the Zócolo (i.e., La Avenida de los Insurgentes)? Would it also be a continuation of the affluent spine? Why or why not?

Keywords: affluent spine, in situ accretion, squatter slums, Mexico City, Griffin-Ford model

Blouet and Blouet, *Latin America and the Caribbean*, 3e
Chapter 7: Mining, Manufacturing, and Services
The Tin Men and Women: Mining in Bolivia

Since the 1860s, Bolivia has been a major source of tin. The story of tin and other mineral extraction in Bolivia is as much a geologic story as it is one of politics, economics, and culture. Let's dig deep into Bolivian mining beginning with a general overview of the country's geography and economy. Create a map of Bolivia using the *Find* command. Click on *Land and Climate* for a quick overview of Bolivia's physical geography. Next, select *Fact and Figures* and then *Society* to read the *Infrastructure* section.

1. Based on what you have read, what basic factors have contributed to the chronic economic development problems in Bolivia?

2. How have natural resources been exploited in the country?

3. What types of minerals are extracted and where are the primary markets?

Now let's take a closer look at the mining industry in Bolivia. Go to *Web Links* and click on "*Bolivia---A Country Study*." This site has an extensive collection of information on all aspects of Bolivia. To focus our attention on the mining industry, use the *Search* tool at the top of the web page and enter "*mining*." Your search results should appear as a list of dozens of links the first of which is titled "*Mining*." Click on "*Mining*" and view the two slides and read the accompanying material.

4. How would you describe the miners depicted in the photographs?

5. What are they extracting and what type of technology is being employed?

Now go back to the mining page and select "*Structure of the Mining Industry*."

6. What is "Comibol" and why was it created?

7. How were "medium" and "small" miners along with cooperatives and others, able to out-produce Comibol by the late 1980s?

Return to the mining page and click on "*Tin and Related Metals*."

8. After reading this section, how would you describe the changes in the Bolivian tin industry in the past 100 years?

9. What types of structural problems within the tin industry exerted political pressure on the Bolivian government?

10. How did the government respond to the collapse of tin prices?

11. Which minerals have supplanted tin as the country's main mining focus?

12. Where are the markets for Bolivian minerals?

Finally, go back to the mining page and from the *Table of Contents* select "*The Liberal Party and the Rise of Tin*" for a more in-depth discussion of mining and politics. **Review** this section and answer the following questions.

13. How did the rise of the Liberal Party relate to the tin industry?

14. Why did the Liberals think it was a sweet idea to move the capital to La Paz?

15. How did these same Liberals intervene in boundary disputes between Bolivia and its neighbors?

16. How did the development of transportation infrastructure by the Liberals and conditions in Europe affect the tin industry?

17. How were indigenous peasants affected by the mining industry?

18. What do you think the future holds for Bolivia and how is their political and economic stability tied to the mining industry?

Keywords: **Bolivia, mining, tin, politics, Comibol, Liberals, trade**

Chapter 8: Mexico

Yúcatan if You Don't Wear Sunscreen: Tourism in the Land of the Maya

One of the facts that many people find amazing about Mexico and Central America is that there are still living descendants of the ancient Maya living all the way from the Yúcatan peninsula to the Central American countries of Guatemala, Honduras and Belize. There may be as many as four million Maya living in the Yúcatan area alone. The Caribbean coast of Mexico including the Mexican states of Yúcatan and Quintana Roo has become a major tourist destination in the past few decades. Tourism has surpassed petroleum as the largest earner of income in Mexico. In fact, a former president of Mexico was heavily involved in the development of the Caribbean coast and it has become the engine of economic development of the region. The ripple effect of tourism is even felt in the state capital of Mérida, hundreds of miles from the coastal development. Many of the locations along Mexico's Caribbean coast, especially Cancún and Cozumel, have been targeted to US tourists. The tourist experience is a multi-faceted one. In this land of the Maya ruin, do you think that the cultural history of the region is foremost in the mind of the tourist or does it merely provide a scenic backdrop and/or diversionary side trip for the sun and fun pleasure seeker? Let's see by taking a virtual tour of the Cozumel-Cancún area. Click on *Find* and type in '*Cozumel*'.

1. Which of these two popular destinations is on an island?

2. What other island is nearby that is also a tourist destination, albeit mainly for Mexicans themselves?

Now type in '*Cancún*' because there are associated *Web Links* with that place name. Click on the web site entitled *Cancún and Cozumel Index* (which will connect you to the new URL for *MexicoWeb*). As you scroll down you will notice a sidebar choice called *The Maya Caribbean Coast*. Click on it and then scroll down that screen until you see a picture of a Maya pyramid and a web site locator for *Maya Adventure*. Click on that locator and choose *Yúcatan*. Once in the Yucatan, choose the site marked *Old Ruins* (as opposed to New Ruins?!) and click on it. Read what they say about the ancient Maya site of *Chichen Itza*. The text suggests that it was the capital of the Mayan Empire. In fact, it was the capital of the highland Maya, a group whose Classic Period came a bit later than the lowland Maya located further south.

3. What forms the "serpent" Kukulcan on the side of the main pyramid?

4. When does the serpent appear?

5. Is this evidence that the ancient Maya were knowledgeable of astronomy? Why or why not?

Now, using the *'Back'* arrow go back to *Maya Adventure* and select *Quintana Roo*. Once again click on *Old Ruins* and read what the text says about the site of *Tulúm*. The site is built on a cliff. Many archaeologists think that its most famous structure--El Castillo-- was used as a lighthouse to warn nearby canoes and ships about the dangerous shoals and reefs in the waters.

6. What other impressive features are found at Tulúm besides El Castillo?

7. What is the distinction of the Coba site some 24 miles inland from Tulúm?

Nearby are other Maya sites along the coast—especially two with intriguing names *'Xcaret'* and *'Xel-Ha'*. Both are now private commercial developments. A better web site to find out about these two places is one called *Cancún Online*. Let's visit *Xcaret* first (virtually that is). First click on "*What to Do*" and scroll down until you find the "*Parks*" heading. Both *Xcaret* and *Xel-Ha* are listed there. Click on *Xcaret* and then click to "*enter paradise*" (and answer the following questions). Click on the sidebar marked *FAQs* (frequently asked questions) and the following displays along the top of the screen—*Aquatic Activities, Maya World*, and *Xcaret at Night*.

8. What did the Maya call the place?

9. What did the Maya do here?

10. Who or what is Ixchel?

11. Why do you think is meant by "all rivers in the Yúcatan Peninsula are underground"?

12. What were the sinkholes (wells) called by the Maya?   By the Spanish?

13. How should the following text really read: "… were the refugee of their goods" (sic)?

14. What is Ulama and how does it appear to differ from soccer?

15. Besides Maya artifacts and 19 scale-model dioramas of important Maya archaeological sites, what weird thing would you find at the Maya World Museum?

Now let's use our virtual powers to hop on over to *Xel-Ha*. Go back to the *Cancún Online* web site and click on "*What to Do*" and then on "*Parks*" and then choose "*Xel-Ha*".

16. What does the name mean in the language of the Maya?

17. What did the Maya use the place for?

18. Why did the gods allow morals to use this place?

So, did this virtual tour whet your appetite for the Mexican Caribbean coast? Do you think that becoming informed about the rudiments of classic Maya culture would add to the enjoyment of a vacation there? Will the Maya ruins serve as focal point or backdrop to your vacation to the Mexican Caribbean coast?

Keywords: **tourism, Cozumel, Cancún, Tulúm, Maya, Chichen Itza, Yúcatan Peninsula**

Chapter 9: Central America
Roller Costa Rica: Riding the Wave of Prosperity

When we think of a Central American country, we often have a stereotype of a banana republic (not the store in the mall) which destroys native species of plants and animals to make room for export-oriented plantation agriculture. We think of countries that are subject to periodic takeovers by military juntas thus allowing the citizenry little freedom of speech or assembly.

We are unlikely to think about a country with a growing and prosperous middle class. Or one with a stable government that is the envy of other third world countries in the region; one that has initiated many protections for endangered species of flora and fauna. Welcome to Costa Rica. It is indeed a rich coast as the name implies in Spanish and a rich contrast to some of its more beleaguered neighbors (e.g., Guatemala, El Salvador, Nicaragua) that have suffered and still suffer the atrocities of civil strife and violence. It was former Costa Rican President Oscar Arias Sanchez who in fact brokered the end to the civil strife in Nicaragua and by so doing won the Nobel Peace Prize in 1987. Transitions of government, mandated every four years, have been remarkably peaceful even when the party in power loses the next election. Since 1948, there have been eleven peaceful transfers of power, a remarkable accomplishment in this part of the world. You can probably imagine why, when choosing among Central American countries, many transnational corporations choose to set up shop and do business in this stable and peaceful country. The threat of expropriation of foreign-owned investments and companies is almost nonexistent and Costa Rica has benefited as a result.

Let's learn more about this fascinating country by clicking on *Find* and then typing 'Costa Rica.' The map that is displayed is the *Comprehensive* variety.

1. What physical feature strikes you about Costa Rica?

2. How large is Costa Rica compared to Guatemala?

In order to find out, click on the sidebar marked *Land and Climate*.

3. What are the names of the three mountain ranges that appear to dominate such a large portion of the landscape?

4. Do these mountain ranges contain active volcanoes? If so, name one.

5. What is the name of the most fertile plateau in the country where much of the commercial agriculture takes place?

6. Where within Costa Rica is the average temperature the hottest?

7. Where is it the coolest?

8. When does the rainy season normally begin and how long does it usually last?

Now click on the sidebar entitled *Facts and Figures* for a thumbnail sketch of the country, basic statistics on the health, education and welfare of its citizenry and its economy. The capital, San Jose, accounts for at least 22 percent of the entire country's population.

9. Is San Jose a primate city (i.e., is its population considerably more than twice the next largest city in the country)?

10. How do you think Costa Rica's life expectancies compare with other countries of Central and South America? (HINT: Go to the pull-down called *Features* at the top of the screen and request *Statistics Center*. Then go to *Population* and search for the specific statistic needed)

11. How do they compare with those of the United States?

12. How do you account for this when the average income of the country would place it in the third world category (i.e., $1,774 gross domestic product per capita)?

Costa Rica is derogatorily referred to as a 'banana republic' because the United Fruit Company once controlled its terms of foreign trade.

13. But what is Costa Rica's main export today?

14. Would the country's literacy rate be an attraction to the typical transnational corporation? Why or why not?

Now click on the sidebar entitled *Society* for a closer look at language and customs of the people.

15. Why are the Costa Rican people called Ticos?

16. Do you think that Costa Rica's high marriage rate has anything to do with the dominant religion of the country? Why or why not?

17. What would you be eating if you were served a national favorite *lengua en salsa*?

18. How about *mondongo*?

19. When might the title Doña be given preference over the title Señora?

20. Why are the beaches especially crowded between January and April?

21. What proportion of the land in this small country has been set aside for nature reserves and national parks?

22. What holiday takes place on April 11th?

23. Who is Rivas?

24. Who is Walker?

25. Who is Juan Santamaria?

26. Who was elected president of the country in 1948 when the current string of stable democratic governments began?

27. Did he serve the country as president after his initial term? If so, when?

28. Where does most of the electricity to supply the industries and cities of the country come from?

29. How does Costa Rica's infant mortality rate of 16 per 1,000 live births, the lowest in Central and South America, compare with the same statistic for the United States?

30. What about the same comparison with Canada?

Now click on the sidebar called *Sights and Sounds*. On the slide entitled "*Rich Farmland in Costa Rica*" what does the caption say are the two large areas of rich volcanic soil where much of the commercial agriculture takes place? Now take a look at "*Costa Rican Coffee Plants*."

31. Where are most of the coffee plants grown? Would Mrs. Olsen (of Folgers' Coffee fame) be proud?

32. If coffee beans are brownish-black in color, why are the berries from which they are derived bright red?

33. Finally, focus on "*Valuable Crop*." Why do you think the bananas are wrapped in plastic?

188

34. What percentage of Costa Rica's workers are farmers?

To learn a bit more about the early history of banana plantations in Costa Rica, click onto the sidebar marked *Web Links* and then on the link called *Encarta Online—Costa Rica*. Now scroll down that material to the section marked "*Agriculture*."

35. In what part of Costa Rica was the world's largest banana plantation developed?

36. What company developed it?

37. What two port cities were developed by that company to ship bananas to world markets? Why do you think that bananas are not raised in the rich and fertile highlands region of the country?

Keywords: **Costa Rica, bananas, coffee, democracy, stable government**

Chapter 10: The West Indies

Volcanoes in the Lesser Antilles: Hot Times in the Islands

The Caribbean. Mention that geographic location and images of sunny vacations, secluded beaches, cultural diversity, and exotic settings spring to mind. What should also appear in your mental slide show are the inherent hazards of living in the Caribbean realm such as hurricanes and volcanoes. Hurricanes come and go and strike mercurially throughout the region but volcanoes are more of a permanent fixture on the landscape. Volcanoes formed many islands in the Caribbean and continue their work of creation and destruction. Situated on the Caribbean Plate, tectonic activity in the region continues to shape the physical geography and daily lives of island dwellers.

To explore the breathtakingly beautiful but equally deadly volcanic landscapes of the Caribbean, begin by creating a map of *Guadeloupe*, site of the first observed eruption in the West Indies. Go to *Web Sites* and click on "*Soufriere Volcano.*" The map on this page illustrates the locations of volcanoes in the *Lesser Antilles* volcanic arc and provides a good reference point for our tour. Take a look at the photo of *La Soufriere* and read the accompanying description. Keeping in mind the conditions in Guadeloupe, go to the "*Volcano World*" link at the bottom of the page and click on the icon. When that web page comes up, use the search engine and enter "*Caribbean.*" The results of your search should yield a list of volcano web sites in the Caribbean. From this list you will be able to take a virtual tour of volcanoes in the region. Select each of the following sites and answer the corresponding questions:

"Dominica, West Indies Location"

1. How do you think the "Valley of Desolation" got its name?

2. Use the *Find* command and create a map of *Dominica* and then click on *Land and Climate*. Why does the United Nations consider this island one the most "disaster-prone countries in the world?" Is it because of volcanoes?

"Slide #6: Mt. Pelee, Martinique, Caribbean" and *"Mt. Pelee, Martinique, Caribbean"*

3. Why is this volcano described as "notorious?"

4. How did Mt. Pelee get its name?

5. Why is the city of St. Pierre in such a dangerous location?

6. What is *nuees ardentes*?

"Soufriere, St. Vincent, West Indies Location"

7. Why is *this* "Soufriere" considered to be more dangerous than the volcano of the same name in Guadeloupe?

"Kick- 'em Jenny, West Indies Location"

8. Other than it's unusual name, what makes Kick-em Jenny so different from other volcanoes in the Lesser Antilles arc?

9. Which two tectonic plates are involved in the creation of this volcano?

"An Old Volcano Awakens on Montserrat"

This account was written prior to the major eruptions in 1997 that brought death and destruction to residents surrounding the Soufriere Hills in Montserrat. For a look at Montserrat after the major eruptions occurred, go back to Encarta and use the *Find* command to create a map of *Montserrat*. Click on *Web Links* and select "*Soufriere Hills Volcano*" and "*Volcano World: Soufriere Hills*." On the "*Soufriere Hills Volcano*" page, be sure to click on "*Government of Montserrat Montserrat Volcano Observatory*" as well as the links for images.

9. What types of landform features have been formed as a result of the Soufriere Hills eruption?

10. How has the British government responded to the emergency situation on Montserrat?

11. How did Hurricane Erika and the volcano combine to form additional hazards for the region?

12. What has happened to the town of Plymouth as a result of the eruptions?

For each of the previously mentioned locations, you can expand your tour by creating a map of each island and selecting *Land and Climate* and *Sights and Sounds*. There are mountains of information that flow out of these sources.

Keywords: **Caribbean, Lesser Antilles, volcanic arc, plate tectonics, volcanoes, hazards, Guadeloupe, Dominica, Martinique, Montserrat, nuees ardentes**

Chapter 11: Andean America

Mountains of Potatoes: The Nature and Cultures of the Andes

The Andes Mountains extend for over 5,000 miles from the Caribbean in the north to Tierra del Fuego at the southernmost tip of the South American continent. Along its course, the Andes cut across cultural, physical, and political regions, creating a barrier between the narrow Pacific coastal region and the bulk of the continent. We will take a general review of several countries within the Andes region including Colombia, Ecuador, Peru, and Bolivia, and discuss their differences and similarities. Then we will change the scale of our investigation by making a side-trip to the Altiplano and Lake Titicaca.

Begin by creating a map of the *Andes Mountains* by using the *Find* command. Read the description of the physical environment by clicking on *Geography* and then take a look at the slides in *Sights and Sounds*.

1. What are some of the challenges faced by those who live in the Andes?

For more detailed information on Andean countries, switch to *Web Links* associated with the *Andes Mountains* and click on "*A Guide to the Andean Countries*." Check out each of the links under "*Contents*" and read the short narratives for *Colombia, Ecuador, Peru, and Bolivia*.

2. After reading about the physical environment, what are some of the more unique aspects of the Andes?

3. How has elevation played a role in the diversity of Andean flora and fauna?

4. How did the colonial period impact indigenous peoples of the Andes?

5. How did the demand for labor and migration during the colonial era influence disease diffusion?

Now that you have read a little about each of the four selected Andean countries, see if you can answer the following questions:

6. Which is the only South American country with ports on both the Pacific Ocean and Caribbean Sea?

7. Which is the smallest of the Andean countries?

8. Which Andean country has the highest percentage of indigenous populations?

9. Which of the Andean countries is largest in area and borders all other Andean states?

10. Which is the only landlocked Andean country?

11. Which of these countries is the most urban?

12. Which has the highest Gross National Product?

Now that you have some background information on the Andes region, let's take a closer look at one location in particular, Lake Titicaca. Go to a map of *Lake Titicaca* (Lago Titicaca) using the *Find* command. Click on *Geography*. How does the lake influence the local climate? How has Lake Titicaca influenced local cultures? You can view an example of local culture by selecting *Sights and Sounds* and viewing the slide titled, "*Uru of Lago Titicaca.*"

Now go to *Web Links* and click on the "*Aymara Culture.*" Read through this material. You can learn more about these indigenous peoples by using the *Find* command to create a map of *Bolivia* and selecting the *Web Link* titled "*Bolivia—A Country Study* ." Then click on the section entitled "*Ethnic Groups*" and read specifically "*Altiplano, Yungas, and Valley Indians.*" Read also the portion entitled "*Aymara, Central*" within the web site *called "Ethnologue Database—Bolivia".*

13. Based on what you have read in these three web sites, what do you think have been the most important historical events that changed the Aymaran way of life?

14. What crops are generally raised in the Lake Titicaca region and why has the potato figured so prominently in this region?

15. What does the future hold for the Aymara and other indigenous groups in the Andes?

Keywords: **Andes Mountains, Colombia, Ecuador, Peru, Bolivia, Lake Titicaca, Aymara**

Chapter 12: Brazil

Were They Brazil Nuts? Brasilia in Retrospect

It's been said that the history of the United States can be summed up in its frontier mentality—that strong desire (or manifest destiny if you prefer) to establish the country from sea to shining sea. Depending on which historian you read, that task was accomplished by the end of the 19th century or certainly by the first decade of the 20th century.

Brazil's sense of manifest destiny is much more recent. Theirs is a great push inward from the large and important coastal cities of Rio de Janeiro and São Paulo into the rich and largely untapped interior of the Mato Grosso and Amazon regions.

One of the symbols of this inward turning was the movement of the capital city from the coast and Rio de Janeiro to Brasilia some 750 miles into the interior of the country. At first, members of the Brazilian parliament, used to the nightlife of cosmopolitan Rio, were reticent to stay in Brasilia on the weekends. Once the conduct of government business was over for the week, the planes would load up with passengers heading to the coast for the weekend leaving Brasilia a bit of a ghost town. Let's find out more about Brasilia and how it might have changed since those early days as the new capital. From the *Find* command, type "*Brasilia*". Study the relative location of Brasilia and then click on the sidebar entitled *Geography* and answer the following questions.

1. What is the approximate population of Brasilia?

2. Is the city located in a tropical rainforest?

3. Is it located in the Amazon region?

4. When did Brasilia become the capital of Brazil?

5. Who was the main force behind the move to the interior that had been suggested ever since 1789?

6. What were the primary reasons for siting the capital and the federal district (Districo Federal) where they are?

7. What architect developed the overall plan for the city of Brasilia?

8. What is Brasilia's layout said to resemble?

9. Who designed most of the public buildings?

10. What large river runs nearby the city?

Now turn to *Sights and Sounds* and examine the caption and slide called *"Modern Architecture in Brasilia"*.

11. How would you describe the setting for the formal government buildings? Does the city look lively and inviting? Why or why not?

Let's explore the architecture of the city in greater detail by clicking on the *Web Links* and going to the site entitled *"Architecture of Brasilia."*

12. From what existing state or states was the Districto Federal carved out?

13. Who was the landscape architect who designed the green spaces, parks and formal gardens?

14. Look at the image of the *Ministerial Esplanade*. What is your impression of that "public space"?

15. Now look at the *Metropolitan Cathedral*. What is the shape of the building supposed to represent?

16. What do you suppose the statuary in front of the building symbolizes?

17. Finally, take a look at *"A Plan of the City"*. Is the plane that the grand plan is supposed to resemble readily apparent?

Now let's explore another of the *Web Links* for Brasilia. Click on the site entitled *"Brasilia's Homepage."*

18. According to the information provided there, what was to be the maximum population of the city according to the planners who developed the master plan (Plan Piloto)?

19. Are the satellite communities needed to house the workers outside of the area covered by the master plan really favelas (squatter slums)? Why or why not?

20. What are the four different meanings attached to the term 'Brasilia'?

21. Using the most generous definition, how many people are there in Brasilia?

22. What is wrong with the myth that the city is completely automobile-oriented, a city of wide-open spaces and broad streets?

23. Can you briefly summarize the status of Brasilia as the capital city around 1964, the time of the military coup and four years after it had been officially declared the capital?

24. Was there a real movement afoot to move the capital back to Rio?

25. Finally, do the legislators still spend as little time as possible in Brasilia, returning instead the 741 air miles to Rio at every opportunity? What would you do if you were a Brazilian politician?

Keywords: **Brasilia, Rio de Janeiro, Brazil, President Juscelino Kubitschek**

Chapter 13: The Southern Cone

Don't Cry for Me Argentina: You Can't be Pompous about the Pampas

The rich grasslands of the Pampas of Argentina are the home of the South American cowboy called the gaucho. There are colorful stories of gauchos not only rounding up and branding cattle, but also capturing the swift and large flightless birds of the Pampas called the rhea with their bolos. The bolo is best described as a cross between a lasso and a slingshot. Wooden balls encased in leather are used for weights and the bola is thrown at the feet of the rhea (or more likely the calf about to be branded) to wrap around them and trip the animal to the ground. Unlike the lasso of the cowboy, once the bola of the gaucho has been released, it must hit its target or be retrieved. Afterwards, the gauchos sit around the campfire sipping yerba maté, a kind of tea popular in the region, through silver sipping straws placed into gourds that serve as cups. At least that is the romantic image of the gaucho and this rich grassland region. What is the reality? Let's find out by exploring Encarta's Virtual Globe '99. Click on the *Find* command and type in *"Pampas, Argentina"*.

1. What does the term pampas mean?

2. From what language is the phrase derived?

3. What are the two parts of the Pampas?

4. Can you think of any other grassland areas that might be similarly divided between humid and semi-arid?

5. What about the Corn Belt of the United States (the humid grassland) and the wheat country of the Great Plains (the semi-arid grassland)? How apt is that analogy?

6. Are the Pampas of Argentina the only pampas in South America?

Now click on *Sights and Sounds* and bring up the image entitled *"Herding Cattle on the Pampas"*.

7. Are you disappointed to find that the modern-day gaucho is just as likely to use a truck for a cattle roundup as a horse?

Using the *Find* command, type in *"Argentina"* and let's see what else we can find out about this country that contains the Pampas. Click on the sidebar called *Land and Climate* and read the material contained there. The rich and fertile Pampas appear to be a product of erosion and deposition.

8. What type(s) of erosion predominate?

9. What is the source area of the deposited material?

10. The Pampas are very large. They extend almost 1,600 kilometers from north to south. What subtropical physical area lies immediately to the north of the temperate Pampas?

11. If the Pampas are so rich and the metropolitan area of Buenos Aires in so large (containing more than ten million of the country's 32+ million people), why is Argentina one of the least densely populated countries in the world?

12. Why do you suppose that per capita agricultural production has declined since 1980 despite very slow population growth in the country?

Now click on the sidebar marked *Society* for more insights into the Argentine situation.

13. When was divorce legalized in Argentina?

14. Why do you think it came so late in the country's history?

15. What do Argentines eat more of on a per capita basis than any other people on earth?

16. Why would street vendors of food products have a tough time making an economic go of it in Buenos Aires?

17. What is considered the country's national sport?

18. Being a nominally Catholic country, Holy Week events are important holidays. What does Maundy Thursday mean?

19. What event is celebrated the day after April Fool's Day?

20. What do Argentines call the Falkland Islands?

21. What does the feast of the Immaculate Conception on December 8th celebrate?

22. At what age is a girl's childhood said to end?

23. How is this rite of passage to womanhood celebrated?

A new form of singing evolved from the folk ballad and nationalistic and political themes.

24. 1.What is this form of music called?

For a sample of what such a song sounds like, click on *Sights and Sounds* and then on the slide entitled "*Historic Plaza.*"

25. Who is the singer?

26. What are the origins of Argentina's national dance—the tango?

27. Would you classify the origins as "highbrow"?

28. What famous president of Argentina was overthrown in 1955 only to be returned from exile and to political power in 1973?

29. What was the name of his second wife (Hint: look at the title of this unit and think of Madonna or Patty Lupone)?

30. This president's third wife was declared Vice President in 1973 and succeeded him as President when he died the next year. Who was she?

31. What was the name of the Marxist guerrilla uprising that prompted the military to take over the government in 1976?

32. What is the "Dirty War"?

33. Who are the "disappeared"?

34. Who is now the president of Argentina and how many times has he been re-elected since first coming to office in 1991?

199

35. What is his family's ethnic background?

36. Who was the chief architect of economic reform that brought inflation down from its annual growth of 4000 percent?

37. When Argentina once again suffered recession and high unemployment in 1996, what happened to Dominic Cavallo?

38. If you worked in Argentina do you think you would enjoy an *aquinaldo*?

Keywords: **Pampas, gaucho, Argentina, Perón, Falkland Islands (Islas Malvinas)**

Chapter 14: Latin America and the World Scene

Banking on the Future of Latin America: Solving Rural Poverty

One of the major challenges facing Latin America at the close of the 20th century is balancing urban and rural development. Many cities such as São Paulo, Brazil, and Mexico City, Mexico, have experienced explosive growth in the post-World War II years but with serious social and economic consequences. A substantial portion of the growing urban populations originated from the rural hinterlands and generally lack marketable skills and financial resources. What are some of the consequences of such uneven growth? The World Bank has examined the rural development issue and considers it to be a serious challenge to Latin American stability as well as a factor that inhibits the region's participation in the global economy.

You can read about Latin America through the eyes of the World Bank by accessing their web site. To do this, create a map of *South America* and select *Web Links* and click on "*The World Bank Latin American and Caribbean*." Focus for a moment on the homepage itself.

1. What appears to be the World Bank's ultimate goal? (HINT: look at the statement in front of their logo at the very top right of the screen)

2. Why isn't having the "highest average per capita income of all the developing world" enough for the countries of Latin America and the Caribbean (LAC)?

Now click on the topic of "*Addressing Rural Poverty*" from "*Major Themes*" sidebar category. After reading that material, then click on "*Rural Development*" located in the sidebar entitled "*Key Sectors*" and answer the following question:

3. What are the World Bank's three major themes for development in Latin America and the Caribbean? (HINT: See the sidebar topics that you can click on for further information)

Read carefully the material labeled "*Rural Poverty in Latin America and the Caribbean*" and click where it says "to learn more" to reveal the full text. Then, answer the following questions.

4. Why is poverty in Latin America often generalized as an "urban problem?"

5. What does the future hold for Latin American "smallholders"?

6. How do ethnicity, education and household characteristics influence economic status in rural areas?

7. Why does the World Bank interpret rural poverty in Latin America differently from other world regions?

8. How have the legacies of colonialism and the plantation economy translated into rural poverty for indigenous peoples today?

9. Why have governments failed to address adequately poverty associated with subsistence agriculture?

10. What is the most important source of cash income for smallholders in most of Latin America at present?

11. Why does the World Bank, an organization once known for large glitzy development projects like hydroelectric dams and iron and steel blast furnaces employing thousands of workers, now frown upon large government programs aimed at relieving poverty? Why are they concentrating their effort and capital on smaller, regionally distributed projects with less areally extensive impacts?

12. How have tariffs on imports and exports as well as internal tax structures affected economic development in rural areas of Latin America?

13. Is investment in "social overhead capital" (e.g., education, health care) detracting attention from agricultural investment in an overall comprehensive strategy of rural development? Why or why not?

Rural and urban poverty in Latin America is widespread and difficult to mitigate. While government programs vary from country to country, the overall picture remains frustratingly constant---a continuation of rural to urban migration, lack of government focus on poverty issues, especially issue of rural poverty, and issues related to land tenure inequality continue to complicate regional economic stability. To gauge the latest thinking of the World Bank about Latin American rural development problems, click on the "*Development Topics*" arrow at the top of the page and when the "*Topics and Sectors*" page comes up, click on "*Rural Development and Agriculture.*" You will be shown a recent Bank publication called *Rural Development: From Vision to Action.* Click on "*Executive Summary*" (which downloads best using Adobe Acrobat reader) and skim the first few pages of the 20 page summary and read carefully beginning on p. 12 when the Latin America and Caribbean region are specifically addressed).

14. If you were hired by the World Bank to address these issues, what would you choose as your main agenda?

Keywords: **Latin America, World Bank, rural development, poverty, agriculture, land tenure**

The Index

A Volume-by-Volume Cross-referencing Guide

This concluding section is designed to help both instructors and students find related material in books other than the one assigned as a textbook for the course. We hope you will find this cross-referencing effort useful. We assume that you will be examining, if not always assigning, every activity associated with the textbook that you have adopted. Since these activities are designed to stand alone from the textbooks and all use *Encarta Virtual Globe '99* CD-ROMs, an instructor can readily use an activity that is not keyed to the particular textbook that he or she is using. These activities can be used for a variety of purposes and in a variety of situations.

For each volume there is a short narrative of our rationale for cross-referencing (if at all appropriate). Some of the activities do not lend themselves well for use in another context but there is a remarkably high degree of transferability in our opinion. At the end of many volume chapters (or parts) of the nine Wiley textbooks examined herein, there is a short statement suggesting an activity (or, in some case, multiple activities) in other chapters of other volumes that also focus on a related concept or region. The cross-referencing of materials is meant to save the instructor time and maximize the degree of attention that can be devoted to a particular subject, concept, country, or region.

For example, if an instructor's focus is on central place theory, there are two activities that specifically address that topic. The nature of the settlement system in the classic central place testing ground of Iowa is the activity developed to accompany "Chapter 9: Take Me Out to the Ball Game: Market Areas and the Urban Hierarchy" in the new interactive human geography book written by Kuby, et al. entitled *Human Geography in Action* (hereafter referred to as Kuby). For those wishing to focus on the more practical and applied aspects of city distribution and central place theory, there is the measurement of actual distances among central places in the Zuider Zee reclamation area of the Netherlands. That activity appears in "Chapter 8: The Location of Tertiary Activities" in Wheeler, *et al. Economic Geography*, third edition (hereafter referred to as Wheeler). Activities that can be cross-referenced give the instructor more freedom of choice in the supplemental materials that he or she might assign to his or her students.

What follows is a volume-by-volume narrative of the chapters (or parts) of other of the nine Wiley texts that form the basis for this activity guide that might be considered supplemental or complementary to the chapter (or part) listed.

VOLUME – Harm J. de Blij and Peter O. Muller, *Geography: Realms, Regions and Concepts 2000*, Ninth Edition (New York: John Wiley and Sons, 2000). Hereafter referred to as 'de Blij and Muller'.

Introduction in de Blij and Muller "World Regional Geography". The focus of the activity *(Back in the USSR: Georgia on My Mind)* is on a trouble spot in the Transcaucasus where bitter ethnic and religious tensions are evident. Likewise, the activity that accompanies **Part Eight of de Blij** ("Geography of Religion") focuses on the historical development and background to the present-day adherence to various Christian denominations in central and eastern Europe *(Christianity Worldwide: Plain Vanilla or 31*

Flavors?). **Part Six in de Blij** ("Landscape and the Geography of Culture") presents Ellsworth Huntington's "beating heart of Asia" hypothesis that was an influential, albeit simplistic, explanation of the troubles in the former Soviet Union.

Chapter 1 in de Blij and Muller "Europe". The activity focuses on the classical theory of agricultural land use put forward by von Thünen and its more contemporary macro-Thünen derivatives *(von Thünen on Whole Wheat? Contemporary European Agricultural Land Use")*. **Chapter 13 in Wheeler** "The Spatial Organization of Agriculture" also focuses on this same important subject. In this case, the activity examines the spatial distribution of large-scale commercial farming in the United States *('Real' Agriculture: Implosion or Explosion?)*.

Chapter 2 in de Blij and Muller "Russia". The purpose of the activity associated with this chapter *(Siberia: Many Rivers Run Through It)* is to provide the student with a virtual field trip through this expansive and often desolate landscape. Likewise, **Exercise 1 in the Physical** triad ("The Earth as a Rotating Planet") includes an activity that examines time zones around the world with particular emphasis on the vastness of Siberia and the many time zones it includes *(Zoning in on Greenwich: Giving You the Time of Day)*.

Chapter 3 in de Blij and Muller "North America" is quite broad in scope. The activity chosen focuses on two separatist movements in culturally pluralistic Canada *(Uh Oh, Canada: A Fractured Federal Tale)*. For a more in-depth examination of Canada's long-term climate change the student might be referred to the activity that accompanies **Exercise 2 in the Physical** books "The Global Energy System" *(There's a Hole, There's a Hole, There's a Hole at the Bottom of the World: Ozone Depletion in Antarctica")*. In addition, Canada's place within Garreau's *Nine Nations of North America* is explored in the activity that accompanies **Chapter 2 of Kuby** *(Postcards from Encarta: Wish You Were Here!)*.

Chapter 4 in de Blij and Muller "Middle America" examines many collisions but the one of interest in the activity is the physical collision of crustal plates that cause seismic activity throughout the region *(The Land of Shake and Bake: Landscapes of Earthquakes and Volcanoes in Mexico)*. The theory of continental drift is explored more in depth in the activity that accompanies **Exercise 9 of the Physical** books "Lithosphere and Tectonics" *(Continental Blue Plate Special: Fetuccine Alfredo Wegener)*. Also, **Exercise 8 of the Physical** books *(Don't Take All Rocks for Granite, It's Sedimentary, My Dear Student, or I Never Metamorphic I Didn't Like)* discusses geologic processes other than tectonic activity. Finally, the activity designed to accompany **Chapter 6 of Blouet** *(Show Some Spine! Land Use in the Latin American City)* touches on the depth-to-bedrock of the sedimentary deposits found throughout the Basin of Mexico.

Chapter 5 in de Blij and Muller "South America". The topic of the activity is a virtual fieldtrip to the rainforest of Suriname and the development potential of such a fragile ecosystem. Preserving the rainforest, distributing land holdings more equitably and alleviating poverty in urban and rural Latin America are the subjects of the activity meant to accompany **Chapter 14 in Blouet** "Latin America and the World Scene" *(Banking on the Future of Latin America: Solving Rural Poverty)*.

Chapter 6 in de Blij and Muller "North Africa/Southwest Asia". The activity examines the plight of a stateless nation (the Kurds) in the Middle East. The political is, of course, but one of many facets of life in the Middle East. **Part 2 of de Blij** "Population and Space" deals with the vast difference between arithmetic and physiologic density in Egypt and other Middle Eastern nations that contain vast amounts of desert *(You Are My Density)*.

Chapter 7 in de Blij and Muller "Subsaharan Africa". This exercise is a sobering one on the "slow plague"—the spread of AIDS in the region.   Two other activities that do focus on Subsaharan Africa include that found in **Part 7 of de Blij** "Patterns of Language" *(You Are What You Speak: African Lingua Francae)* and that in **Chapter 11 of Wheeler** "Manufacturing: Regional Patterns and Problems" which focuses, in part, on Angola *(Diamonds in the Rough: Facets of Industrial Development)*. **Exercise 17** accompanying the **Physical** books *(Our Just Deserts: Soil Erosion)* explores an aspect of the physical environment that is exacerbating the spread of this infectious killer.

Chapter 9 in de Blij and Muller "East Asia". The activity accompanying this chapter is a focus on the burgeoning city and enterprise zone of Shanghai. A good complement to this urban focus is the activity associated with **Exercise 15 in the Physical** books "The Work of Waves and Wind" . The activity *(Loess is More)* examines the fertility of the loessal plains of China and their importance for Chinese agriculture both historically and currently.

Chapter 10 in de Blij and Muller "Southeast Asia". The activity associated with this chapter has to do with the beautiful, backward, and inward-looking country of Mynamar (formerly Burma). Another theme that might be pursued in Southeast Asia (and beyond) is the influence of the Chinese Diaspora on the economy of the entire region. **Part 12 of de Blij** "The Political Imprint" contains an activity that focuses specifically on the overseas Chinese who escaped from Hong Kong before the Chinese takeover in July 1997 *(The Yacht Sea People: Roll the Dice and End Up in Vancouver)*.

Chapter 11 in de Blij and Muller "The Austral Realm". Australia is also alluded to in an activity to accompany **Chapter 3 of Kuby** "Tracking the AIDS Epidemic: Diffusion Through Time and Space". One of the three Chinatowns discussed in *Digging to Chinatown: Relocation Diffusion in Action* is that of Sydney, Australia.

Chapter 12 in de Blij and Muller "The Pacific Realm". Another activity that focuses on Micronesia and Fijian culture is found in **Chapter 5 of the Physical** texts "Winds and the Global Circulation System". The activity *(A World Wind Tour)* focuses on the importance of sailing winds to the peoples of the South Pacific. Likewise, the activity accompanying **Part Four of the de Blij** volume *(Not So Great (Life) Expectations)* focuses on demographic issues throughout the world, many of which impact the Pacific Realm.

VOLUME – Harm J. de Blij, *Human Geography: Culture, Society, and Space*, Fifth Edition, (New York: John Wiley and Sons, Inc., 1996) and Harm J. de Blij and Alexander B. Murphy, *Human Geography: Culture, Society and Space*, Sixth Edition (New York: John Wiley and Sons, Inc., 1999). Hereafter referred to as 'de Blij'.

Part 1 in de Blij "Environment and Humanity" is a broad subject area. The activity focuses on a small section of the chapter dealing with revealed residential desirability via mental mapping *(Rocky Mountain High (and Corn Belt Low))*. A broader perspective might be found in **Chapter 5 of Kuby** "Trapped in Space: Space-Time Prisms and Individual Activity Space". The accompanying activity *(My Prism can be my Prison)* allows the student to work with the concepts of time-geography as discussed by Törsten Hägerstränd and his associates. Also related is **Chapter 1 of Kuby** "True Maps, False Impressions:

Making, Manipulating and Interpreting Maps" with its activity on the concept of geographic scale.

Part 2 in de Blij "Population and Space" is another multi-faceted subject area. The activity in the book focuses on differing measures of population density *(You are my Density)*. Another important aspect of population is the age-sex cohort (i.e., the population "pyramid") which is the focus of the activity in **Chapter 7 of Kuby** "The Hidden Momentum of Population Growth". The activity focuses specifically on Egypt *(Another Type of Egyptian Pyramid)*.

Part 3 in de Blij "Streams of Human Mobility" is an examination of migration throughout the world. The activity that accompanies the chapter *(Like Salmon Swimming Upstream?)* focuses on African-American migration (and subsequent counter migration) in the United States. For a broader perspective, you might also use the activity that accompanies **Chapter 5 in Blouet** "Population: Growth, Distribution, and Migration" entitled *Why you have to go to New York City to get your Go-Go Boots Shined: Brazilians on the Move* which refers to the most common job titles held by Brazilians migrating to New York City.

Part 4 in de Blij "Patterns of Nutrition and Health" can be tied into the activity that accompanies **Wheeler Chapter 2** "Global Population Processes and Pressures" entitled *Painting the World by Numbers*. An entire series of demographic and public health variables are examined.

Part 5 in de Blij "Geography and Inequality" can be reinforced by assigning the activity contained in **Kuby Chapter 12** (*Timor or Less: Indonesia's Claim at Stake*) that focuses in part on gender issues in beleaguered Timor.

Part 6 in de Blij "Landscape and the Geography of Culture" can be embellished by having students examine the activities associated with the **Introductory chapter in de Blij and Muller** ("Back in the USSR: Georgia on my Mind"). Both activities focus at least in part on the nature of ethnically diverse and fragmented areas such as the Transcaucasus and the Balkans. Also, the activity in **Exercise 7 of the Physical** texts (*Köppen, Schmerpen: How Can Botany Possibly Aid Climatology?*) may help students to understand the difference between climatic influences on human behavior and belief in environmental determinism

Part 7 in de Blij "Patterns of Language" is a fascinating subject. The activity in de Blij focuses on the language of business and trade in Africa (*You Are What you Speak: African Lingua Francae*). The focus of **Chapter 12 in de Blij and Muller** is on languages too, but this time those of the South Pacific (*Want to Know Samoa? Read On!*).

Part 8 in de Blij "Geography of Religion" is quite broad ranging. The activity to accompany this chapter is more directly focused on the variability of religious expression within the rubric of Christianity. Even more focused is the activity to accompany **Chapter 11 in Kuby** "Do Orange and Green Clash? Residential Segregation in Northern Ireland" on the religious conflicts between Catholic and Protestant in Northern Ireland (*I'll see your Bernadette Devlin and Raise you an Ian Paisley*). The subject of religion and ethnicity in the Caucasus Mountain region of the former Soviet Union is the subject of the **Introductory Chapter of de Blij and Muller** (*Back in the USSR: Georgia on my Mind*) and that activity might also be used.

Part 9 in de Blij "Cultural Landscapes of Farming" can encompass anything from subsistence farming by small-plot cultivators to commercial agro-business enterprises on a grand scale. The activity in de Blij *(Let them eat Jute: The Colonial Legacy in World Agriculture)* focuses on peasant farmers in Bangladesh. On the other hand, the activity that accompanies **Chapter 13 in Wheeler** "The Spatial Organization of Agriculture" *('Real' Agriculture: Implosion or Explosion?)* focuses on modern commercial agriculture in the United States.

Part 10 in de Blij "The Urbanizing World" can be related to other activities in other books including those which accompany **Kuby Chapter 9** *(Towns in Iowa: Central Places and a Whole Lot More!)* and **Birdsall Chapter 4** *(Urban Behemoths: Wave of the Future or Relicts of the Past?)*

Part 11 in de Blij "Cultures, Landscapes, and Regions of Industry" contains an activity that focuses mainly on the maquiladoras in the Tijuana-San Diego area. For completeness of coverage, the instructor is encouraged to consider assigning the activity accompanying **Chapter 10 in Wheeler** "Manufacturing: Where Plants Locate and Why" as well *(Run for the Border: NAFTA)*. Likewise, the activity accompanying **Chapter 14 in Blouet** *(Banking on the Future of Latin America: Solving Rural Poverty)* is related to the general subject matter covered.

Part 12 in de Blij "The Political Imprint" discusses the recent Hong Kong Diaspora prior to the takeover of the former British colony by China. Their impact has been especially felt in Vancouver. Likewise, **Chapter 3 in Kuby** "Tracking the AIDS Epidemic: Diffusion Through Time and Space", despite the chapter title, contains an activity that is closely related to that in Part 12 of de Blij. The activity is entitled *Digging to Chinatown: Relocation Diffusion in Action*. One of three Chinatowns that the activity focuses on is Vancouver's Chinatown.

VOLUME – Michael Kuby, John Harner and Patricia Gober, *Human Geography in Action* (New York: John Wiley and Sons, Inc., 1998). Hereafter referred to as 'Kuby'.

Chapter 1 in Kuby "True Maps, False Impressions: Making, Manipulating, and Interpreting Maps" examines the whole of cartography. The accompanying activity *(Large Map, Small Scale; Small Map, Large Scale)* focuses on the concept of map scale whereas the activity designed to accompany **Part 1 of de Blij** "Environment and Humanity" examines the relevant area of mental mapping *(Rocky Mountain High (and Corn Belt Low))*.

Chapter 2 in Kuby "Cactus, Cowboys, and Coyotes: The Southwest Culture Region" examines how cultural stereotypes evolve and are maintained. Maintaining identity in a culturally pluralistic society is the related concept examined in the activity that accompanies **Chapter 3 in de Blij and Muller** "North America". The activity is entitled *Uh Oh, Canada: A Fractured Federal Tale,* discusses how places can be maligned because of a poor public image. The states of Ohio and Iowa, both of which now rank quite low on most students' mental maps of the United States, are explored.

Chapter 3 in Kuby "Tracking the AIDS Epidemic: Diffusion Through Space and Time" examines the epidemiology of our modern scourge on humanity. Migrations can also be studied using the same diffusion mechanisms and students might wish to extend the overseas Chinese focus of the activity to other realms. The exercise that accompanies **Chapter 5 in**

Blouet "Population: Growth, Distribution, and Migration" discusses recent Brazilian migrations into New York City *(Why you have to go to New York City to get your Go-Go Boots Shined: Brazilians on the Move)*. Finally, the activity accompanying **Part 12 in de Blij** *(The Yacht Sea People: Roll the Dice and End up in Vancouver)* addresses the outmigration of relatively wealthy Hong Kong residents to Vancouver before the 1997 takeover of the colony by China.

Chapter 4 in Kuby "Newton's First Law of Migration: The Gravity Model" extols the virtues of natural laws to explain aggregate behavior. If the instructor wishes to emphasize other applications of geographic location theory, he or she might wish to assign the activity that accompanies **Chapter 8 of Wheeler** "The Location of Tertiary Activities". The activity is entitled *Here's the Rub: The Dutch have to Modify Central Place Spacing*. Biases such as the local dome effect that might affect the shape and nature of the gravity formulation are addressed tangentially in the activity developed to accompany **Part 1 of de Blij** *(Rocky Mountain High (and Corn Belt Low))*.

Chapter 5 in Kuby "Trapped in Space: Space-Time Prisms and Individual Activity Space" should have students wondering about the role of transportation in causing the time-space convergence they continue to experience. Another venue that considers this issue is **Chapter 3 in Wheeler** "Global Economic Development". The activity is entitled *Thumbs Down: A Hitchhiker's Guide to the Globe*, and it should reinforce the uneven nature of personal transportation availability throughout the world.

Chapter 6 in Kuby "Help Wanted: The Changing Geography of Jobs" contains an activity that compares the economic base of countries a variety of development levels. The country that has the highest percentage of employment in manufacturing is Angola. The incredulous student (and instructor) might wish to pursue this further by examining the activity that accompanies **Chapter 11 in Wheeler** "Manufacturing: Regional Patterns and Problems". The activity entitled *Diamonds in the Rough: Facets of Industrial Development* might indicate why Angola is the world's leading manufacturer (as measured by percentage of the gainfully employed laborforce in that sector). See also the activity in **Chapter 12 of Blouet** *(Were they Brazil Nuts? Brasilia in Retrospect)* which focuses on the manner in which the Newly Industrializing Country (NIC) of Brazil set a course for its future expansion.

Chapter 7 in Kuby "The Hidden Momentum of Population Growth" contains an activity that focuses largely on Egypt *(Another Type of Egyptian Pyramid)*. For the instructor who might wish to intensify this Egyptian focus, we would suggest **Part 2 in de Blij** "Population and Space". The accompanying activity *(You are my Density)* examines the difference between arithmetic and physiologic density that is greatest in Egypt.

Chapter 8 in Kuby "From Rags to Riches: The Dimensions of Development " might be cross-referenced with any number of activities in other books. One that stands out is **Chapter 4 in Wheeler** "The Interdependent Global Economy" with an activity that focuses on banking in a third world context *(Banking on Technology)*. Don't overlook the importance of the physical resource endowment of the countries either. **Exercise 5 in the Physical** books *(A World Wind Tour)* and **Exercise 13 of the Physical** volumes *(Wet and Wild Waterfalls)* point to renewable resources that are just beginning to be tapped by the developing nations of the world. The contrast in development levels is a theme that can, of course, be found in other chapters too. We'd be interested in finding out which of the activities works best for the instructor and resonates most with the students.

Chapter 9 in Kuby "Take Me Out to the Ball Game: Market Areas and the Urban Hierarchy" is a great example of applied location theory especially the elements of central place theory. Hewing closer to the initial theorems of the classical theory is the spacing among settlements in the Netherlands that were reclaimed from the polders and the subject of **Chapter 8 in Wheeler** *(Here's the Rub: The Dutch Have to Modify Central Place Spacing)*. Don't overlook the activity to accompany **Part 10 of de Blij** *(Planet of the Apes? The Primate City Distribution)* or even **Exercise 15 in the Physical** books *(Loess is More)* to explain why it is that some agrarian regions are so agriculturally abundant in the physical sense and, therefore, supportive of central place services.

Chapter 10 in Kuby "Reading the Urban Landscape Through Census Data and Field Observation" addresses an important aspect of urban geography—empirical field work. The activity developed to accompany this chapter *(Manhattan Transformation: The Suburban Roots of Harlem)* has the student thinking backwards about what a central city area today might have been like when it was a suburb on the periphery of an expanding metropolitan region. An activity that has students speculating about the urban future is found in **Chapter 7 of Wheeler** "The City as an Economic Node". The activity has the students focus on Houston--that burgeoning city in Texas best known for having its fortunes tuned to the health of the "oil patch" economy and for having no zoning regulations *(Houston, We Have a Problem: Invasion of the Multiple Nuclei)*. The students should speculate about urban sprawl and land use planning.

Chapter 11 in Kuby "Do Orange and Green Clash? Residential Segregation in Northern Ireland" deals with religiously based conflict in Northern Ireland. The student might also examine the activity in **Part 8 of de Blij** "Geography of Religion" entitled *Christianity Worldwide: Plain Vanilla or 31 Flavors*?
Chapter 12 in Kuby "The Rise of Nationalism and the Fall of Yugoslavia" examines the tragedy of the Balkans. The associated activity looks at another neo-colonial nationalistic clash that receives less attention in North America—the Indonesian claims to Timor *(Timor or Less: Indonesia's Claim at Stake)*. A good companion activity is the one associated with **Part 5 of de Blij** "Geography and Inequality" that focuses on inequality in gender roles around the world *(A Woman's Place is in the House...and in the Senate)*.

Chapter 13 in Kuby "How Would You Like Your Animals—Rare? Vacancies at the World Zoo" examines the sobering subject of global extinction of certain animal species. **Exercise 18 in the Physical** volumes *(Act Locally, Think Globally)* also examines aspects of the impending global environmental crises and might be a good complement.

VOLUME – The triad of Physical Geography textbooks that include: Alan Strahler and Arthur Strahler, *Physical Geography: Science and Systems of the Human Environment* (New York: John Wiley and Sons, Inc., 1997); Alan Strahler and Arthur Strahler, *Introducing Physical Geography*, 2nd Edition (New York: John Wiley and Sons, 1998); and Harm J. de Blij and Peter O. Muller, *Physical Geography of the Global Environment* (New York: John Wiley and Sons, 1997).

Exercise 1 in Physical "The Earth as a Rotating Planet" contains an activity dealing with time zones with an emphasis on the immense size of Russia, especially Siberia *(Zoning in on Greenwich: Giving You the Time of Day)*. Siberia is also discussed in **Chapter 2 of de Blij and Muller** "Russia" in the activity entitled *Siberia: Many Rivers Run Through It*.

Exercise 4 in Physical "Atmospheric Moisture and Precipitation" develops an activity entitled *Somewhere over the Rain Shadow*. The orographic effect is one cause of desert or steppe-like conditions on the rain shadow side. Other mechanisms are discussed in a focus on the Atacama Desert of Chile in **Chapter 2 of Blouet** "Physical Environments of Latin America". The related activity is entitled *Where Not to Be in Raincoat Sales: High and Dry in the Atacama Desert*.

Exercise 5 in Physical "Winds and the Global Circulation System" is built upon an activity that examines wind patterns around the world including the tropics. For a more in-depth look at how the wind and weather patterns in some of these exotic tropical locales affects the tourist trade see **Chapter 8 in Kuby** "From Rags to Riches: The Dimensions of Development". The emphasis of the activity is on the emerging niche market for ecotourism *(Ecotourism: It Isn't Easy Being Green—But it Can be Profitable)*. Similarly, the activity developed to accompany **Chapter 4 in Wheeler** *(Banking on Technology)* discusses the motivation underlying the nonaligned nations movement located, by and large, in the tropical regions where tourism is often the primary source of external revenue.

Exercise 7 in Physical "The Global Scope of Climate; Low-Latitude Climates; Midlatitude and High-Latitude Climates" has an activity that explores the Koeppen classification system. But the relationship between climate and human behavior has been a hotly debated topic for millennia. In the early part of this century the theoretical constructs of environmental determinism mesmerized the field of geography (and many others as well). That "theory" is the focus of the activity representing **Part 6 of de Blij** "Landscape and the Geography of Culture" entitled *What'll it be?: Lethargic and Clever or Vigorous and Stupid?*

Exercise 8 in Physical "Earth Materials and the Cycle of Rock Change" presents the three different classes of rocks and their geological origins. For an application of the geology lesson, students may find **Chapter 4 of de Blij and Muller** "Middle America" of interest. The exercise emphasizes two tectonic processes affecting the landscape of Mexico City *(The Land of Shake and Bake: Landscapes of Earthquakes and Volcanoes in Mexico)*.

Exercise 9 in Physical "Lithosphere and Tectonics" is technical and so is the activity developed for it dealing with the theory of plate tectonics *(Continental Blue Plate Special: Fetuccine Alfredo Wegener)*. If the student wants a more humanistic perspective on how the physical environment created by such tectonic activity affects the daily lives of the people who live in these extreme environments, see **Chapter 11 in Blouet** "Andean America" and the activity entitled *Mountains of Potatoes: The Nature and Cultures of the Andes*.

Exercise 10 in Physical "Volcanic and Tectonic Landforms" does an excellent job of explaining the rudiments of volcanism and the accompanying activity *(Return to Cinder)* is meant to bring that learning closer to home by focusing on an event that every college student in the United States is likely to recall—the 1980 explosion of Mount St. Helens in Washington. Volcanoes, especially when they are inactive, also make for fascinating landscapes as pointed out in the activity developed to accompany **Chapter 10 in Blouet** "The West Indies" entitled *Volcanoes in the Lesser Antilles: Hot Times in the Islands*.

Exercise 13 in Physical "Fluvial Processes and Landforms" covers a variety of water-related topics but spectacular waterfalls hold a special interest for residents and tourists alike and form the basis of the activity *(Wet and Wild Waterfalls)*. Another activity that picks up on the theme of turning natural attractions such as waterfalls into tourist amenities is

Chapter 8 in Kuby on ecotourism. The activity is entitled *Ecotourism: It's Not Easy Being Green--But it can be Profitable.*

Exercise 15 in Physical "Glacier Systems and the Ice Age" focuses on an area in southwestern Wisconsin and portions of Iowa and Illinois that were missed by the most recent glacial period. This so-called Driftless Area forms the basis for the activity *(The Iceman Misseth: Implications for the Driftless Area).* **Chapter 9 in Kuby** "Take Me Out to the Ball Game: Market Areas and the Urban Hierarchy" also focuses on the central place system in Iowa, the northeastern part of which is included in the Driftless Area. That physical factor may partially explain the spacing of settlements in that quadrant of the state, the focus of the chapter activity *(Towns in Iowa: Central Places and a Whole Lot More!).*

Exercise 18 in Physical "Systems and Cycles of the Biosphere and Global Ecosystems" is one of those sweeping catch-all chapters that tries to focus on ecological imbalances planet-wide. The activity designed to go along with this chapter encourages the reader to become proactive about ecological issues of concern to them within their locales *(Act Locally, Think Globally).* Global sustainability, especially the alarming rate of animal and plant extinctions, is also the subject of **Chapter 13 in Kuby** "Human Impact on the Environment". The activity entitled *How Would You Like Your Animals—Rare? Vacancies at the World Zoo* is a good complement to the physical exercise.

VOLUME – James O. Wheeler, Peter O. Muller, Grant Ian Thrall and Timothy J. Fik, *Economic Geography***, Third Edition (New York: John Wiley and Sons, Inc., 1998). Hereafter the volume shall be referred to as 'Wheeler'.**

Chapter 1 in Wheeler "The Study of Economic Geography" is an overview of many concepts used in economic geography. The activity associated with the chapter *(One Picture Tells a Thousand Geographies)* focuses on the book cover photograph which so ironically captures elements of the traditional and modern simultaneously. A related activity might be found in **Chapter 2 of de Blij and Muller** "Russia". That activity *(Siberia: Many Rivers Run Through It)* attempts to put a human face on a region that is supposedly filled with emptiness, loneliness, and depression. The student should come away somewhat disabused of that notion.

Chapter 2 in Wheeler "Global Population Processes and Pressures". The activity for this chapter examines many of the population/demographic statistics available on *Encarta Virtual Globe '99* CD-ROMs. A more focused look at population density can be found in the activity designed to accompany **Part 2 of de Blij** "Population and Space". That activity is entitled *You Are My Density*. **Part Four of the de Blij** volume *(Not So Great (Life) Expectations)* focuses on demographic issues throughout the world, many of which impact the Pacific Realm.

Chapter 3 in Wheeler "Global Economic Development". The activity associated with this chapter examines the vast disparities between those countries that have reasonable access to private means of transportation and those who do not. A good complementary activity is found in **Chapter 5 in Kuby** "Trapped in Space: Space-Time Prisms and Individual Activity Space". It is entitled *My Prism Can be my Prison* and helps to illustrate the concept of the time-space convergence so crucial to an understanding of advances in transportation technology.

Chapter 4 in Wheeler "The Interdependent Global Economy". Global interdependence and global capitalism is an awfully large subject to embrace in a single activity. The activity on banking in Wheeler *(Banking on Technology)* might be complemented by the one found in **Chapter 14 of Blouet** "Latin America and the World Scene" *(Banking on the Future of Latin America: Solving Rural Poverty)* that deals with changing policy toward rural development of the World Bank. Also, the activity developed for **Chapter 8 in Kuby** *(Ecotourism: It Isn't Easy Being Green—But It Can Be Profitable)* broaches one potential source of income that many Latin American countries could tap.

Chapter 5 in Wheeler "Principles of Spatial Interaction". The activity designed to accompany this chapter deals with the stage model of transportation development first discussed by Taaffe, Morrill, and Gould *(Geography is Very Spatial to Me!)*. A good complement to the contemporary focus on transportation systems within different states of the United States is found in the activity associated with **Chapter 3 of Blouet** "Aboriginal and Colonial Geography of Latin America" *(Trail of Years: Historical Linkages Along the Inca Highway)*.

Chapter 6 in Wheeler "The Role of Transportation in Economic Geography". The activity here is an historic look at the relatively short-lived canal building craze in the United States. For a possible complement you might wish to examine an exotic, albeit modern means of transportation aboard the Trans-Siberian Railway by doing the activity suggested in **Chapter 2 of de Blij and Muller** "Russia" *(Siberia: Many Rivers Run Through It)*.

Chapter 7 in Wheeler "The City as an Economic Node". It might be interesting to compare and contrast the urban evolution of Houston, Texas a city that can annex surrounding suburban territory easily and for which zoning is mostly an afterthought with more "typical" suburban development. The activity associated with **Chapter 10 in Kuby** "Reading the Urban Landscape Through Census Data and Field Observation" *(Manhattan Transformation: The Suburban Roots of Harlem)* compares Harlem's transition from early suburb of New York to modern inner-city ethnic neighborhood with King of Prussia, Pennsylvania, that began as a dormitory suburb of Philadelphia and grew into a suburban downtown node of activity.

Chapter 8 in Wheeler "The Location of Tertiary Activities". The activity for this chapter discusses the spacing of settlements developed *de novo* from the reclaimed polderland in the Netherlands. Location theory is not immutable as the Dutch found out when they tried to hew too closely to the Christaller version of the theory thirty years after it was developed. The perfect complement to this activity is the one that accompanies **Chapter 9 in Kuby** "Take Me Out to the Ball Game: Market Areas and the Urban Hierarchy". The activity *(Towns in Iowa: Central Places and a Whole Lot More!)* has the student measure nearest neighbor distances among central places of high order in the state of Iowa, the classical central place testing ground. The results of the comparison with the Netherlands (and thus southern Germany) are startling. Also, **Part 10 of de Blij** "The Urbanizing World" has an activity that classifies the city systems of many countries in the world as primate or rank-size.

Chapter 9 in Wheeler "The Changing Economic Geography of the Restructured Metropolis". The activity accompanying this chapter *(A Tale of Three Cities)* examines New Orleans, Denver, and Baltimore and what they've given up to their surrounding suburbs. A good contrast might be to examine planned communities that did not just grow like topsy with no rhyme or reason. A complement to this might be the post-industrial change going on in two former heavy industrial centers—Cleveland and Pittsburgh. These two cities are featured in **Birdsall Chapter 5** "North America's Manufacturing Core" *(Gritty Cities and Rust Belt or Nexus for a Brighter Future?)*. Both cities have started to make the (sometimes painful) transition from high-wage semi-skilled industrial jobs to either lower wage service jobs (at least temporarily) or more highly skilled job opportunities in a variety of economic sectors.

Chapter 10 in Wheeler "Manufacturing: Where Plants Locate and Why". The activity in this chapter focuses on maquiladora plants in the Cuidad Juarez-El Paso area *(Run for the Border: NAFTA)*. It is fitting therefore that the student might also be assigned to complete the activity associated with **Part 11 of de Blij** "Cultures, Landscapes, and Regions of Industry". That activity *(Maquiladora? Is that the Latest Latin Dance Craze?)* focuses mainly on the San Diego-Tijuana area of the US-Mexican border. In addition, the activity accompanying **Chapter 14 in Blouet** *(Banking on the Future of Latin America: Solving Rural Poverty)* focuses on the seemingly intractable problem of unequal land distribution in Latin America.

Chapter 11 in Wheeler "Manufacturing: Regional Patterns and Problems". The activity here focuses mainly on mining in Angola, the country with the highest proportion of its labor force in manufacturing. Activities in two other chapters are related and might also be assigned. The first is **Chapter 7 in Blouet** "Mining, Manufacturing and Services". The activity here *(Tin Men and Women: Mining in Bolivia)* might be compared to a mineral economy in Angola based on oil and formerly on diamond extraction. The second activity is found in **Chapter 6 of Kuby** "Help Wanted: The Changing Geography of Jobs". The activity *(Around the World from Pre- to Post-Industrial)* has the student examine the employment profile of five disparate countries at various stages on the continuum of economic development.

Chapter 13 in Wheeler "The Spatial Organization of Agriculture". The activity associated with this chapter updates Hart's concept of 'real' agriculture using more recent data from the agricultural census. We recommend assigning the activity associated with **Chapter 1 in de Blij and Muller** "Europe" as well. The activity *(von Thünen on Whole Wheat? Contemporary European Agricultural Land Use)* tests whether macro-Thünen models have any relevance in explaining modern European agricultural land use patterns. Also, **Part 9 of de Blij** contains an activity *(Let Them Eat Jute: The Colonial Legacy in World Agriculture)* that may aid in understanding why acreage that could be used to grow the basic necessities is instead used to produce commercial crops for export.

Chapter 14 in Wheeler "Contemporary American Agriculture: Regions and Trends". The activity in this chapter focuses on the commercial dairy belt in the United States. The instructor may, however, wish to contrast commercial agriculture in the United States, which is really big business, with small plot cultivation in the third world. A good activity for this is **Part 9 of de Blij** "Cultural Landscapes of Farming" entitled *Let Them Eat Jute: The Colonial Legacy in World Agriculture*. This activity examines the disastrous consequences of a saturated world market for a commercial cash crop such as jute, the commercial staple of Bangladesh. Furthermore, the activity developed to accompany **Birdsall, Chapter 11** *(Agricultural Surpluses: Mutiny over the Bounty?)* focuses on soybeans, a major exportable food product supplying the needs of global markets.

VOLUME – Stephen S. Birdsall, John W. Florin and Margo L. Price, Regional Landscapes of The United States and Canada, Fifth Edition (New York: John Wiley and Sons, Inc., 1999). This volume will be referred to as 'Birdsall'

Chapter 1 in Birdsall "Regions and Themes". The activity in this chapter *(Despite Its Many Faults, California is Everything It's Cracked Up to Be)* focuses on the diversity of California and the fact that it can be divided into more subregions than most other states. Complementing the (usually positive) impression of California, **Part 1 in de Blij** ("Environment and Humanity") focuses on two states that are sometimes maligned in mental mapping surveys—Iowa and Ohio *(Rocky Mountain High (and Corn Belt Low))*. The exercise introduces students to the variety and diversity of these archetypal Middle Western states to illustrate that even the Corn Belt can be diverse and interesting. Along the lines of how regionalization should be accomplished is the activity found in **Kuby Chapter 2** ("Cactus, Cowboys, and Coyotes: The Southwest Culture Region"). The activity *(Postcards from Encarta: Wish You Were Here!)* is based on Joel Garreau's best-selling book *The Nine Nations of North America* and should stimulate thinking about how regionalization is accomplished and for what purpose(s). If the interest in California is more of a physical nature, **Exercise 4 in the Physical books** ("Atmospheric Moisture and Precipitation") focuses on the rain shadow effect of tall mountains such as the Sierra Nevadas. Students are invited on a virtual tour of the differences between the windward and leeward sides of such mountain ranges in this activity that employs the Virtual Flights tool of *Encarta Virtual Globe '99 (Somewhere Over the Rain Shadow)*.

Chapter 4 in Birdsall "Megalopolis". This activity focuses upon the fact that there are now many megalopolises throughout the world since Gottman first coined the phrase to refer to the Boston to Washington corridor along the Eastern seaboard of the United States *(Urban Behemoths: Wave of the Future or Relicts of the Past?)*. The activity accompanying **Kuby Chapter 4** ("Newton's First Law of Migration: The Gravity Model") employs the magnifying capabilities of the *Encarta* tool to zoom in on population concentrations around the world. These concentrations are viewed at night with *Encarta* in order to create more visual drama and better see urban agglomeration characteristics *(A Matter of Some Gravity: World Population Cores and Distance Decay)*. Likewise, **Part 10 of de Blij** "The Urbanizing World" has an activity *(Planet of the Apes? The Primate City Distribution)* that explores the systemic properties of city systems in developed and developing countries.

Chapter 5 in Birdsall "North America's Manufacturing Core". This activity *(Gritty Cities and Rust Belt or Nexus for a Brighter Future?)*, focuses on structural changes in the economic base of Pittsburgh and Cleveland--two centers commonly associated with the Rust Belt phenomenon. Likewise, the activity designed to accompany **Chapter 9 in Wheeler** *(A Tale of Three Cities)* focuses on three cities—New Orleans, Denver, and Baltimore—in which their suburban fringe areas have been usurping industry, business, and service functions from the central cities at a rapid rate. Are there lessons to be learned for other cities in the United States?

Chapter 11 in Birdsall "The Agricultural Core". The activity designed for this chapter *(Agricultural Surpluses: Mutiny over the Bounty?)* focuses on two states of the traditional Corn Belt which might more properly be called the Corn-Soybean belt. The lowly soybean with its many uses and useful properties is the focus of attention, especially the extent to which soy products are exported overseas. The activity to accompany **Chapter 14 in Wheeler** *(How Now Dairy Cow?)* simply aims its sights a bit higher (in latitude) at the hay-dairy belt of Wisconsin rather than the corn-soybean belt of Iowa and Illinois. Together

214

these activities should provide a more complete picture of Midwestern agriculture than either standing alone.

Chapter 16 in Birdsall "The North Pacific Coast". The activity designed to accompany this chapter *(This Isn't Tan, It's Rust: Raining on the Emerald City's Parade?),* focuses on the West Coast marine climate as it affects Seattle in particular. For a complementary view of what goes on over on the leeward (rain shadow) side of the Cascades, check out the activity meant to accompany **Chapter 4 in the Physical** textbooks *(Somewhere Over the Rain Shadow).*

VOLUME – Brian W. Blouet and Olwyn M. Blouet, *Latin America and the Caribbean: A Systematic and Regional Survey* **(New York: John Wiley and Sons, Inc., 1997) Hereafter referred to as 'Blouet'.**

Chapter 2 in Blouet "Physical Environments in Latin America" can be linked to **Physical Chapter 4** if a further elaboration of the rain shadow effect in a North American context is desired for contrast with its effect in the Andes.

Chapter 5 in Blouet "Population: Growth, Distribution and Migration" can be linked to **Part 3 of de Blij,** which focuses on the Great (African-American) Migration in the United States, as a comparative study with migration movements in Latin America.

Chapter 6 in Blouet "The Latin American City" can be linked to **de Blij and Muller Chapter 4** "Middle America" (activity entitled *The Land of Shake and Bake: Landscapes of Earthquakes and Volcanoes in Mexico*). This activity focuses specifically on the devastating earthquake in 1985 and why Mexico City is so vulnerable to seismic activity. This complements the material on the urban spatial structure of Mexico City as an illustration of the Griffin-Ford model in the Blouet chapter.
Also related, albeit more tangentially, is **Physical Chapter 10** ("Volcanic and Tectonic Landforms") and its activity entitled *Return to Cinder*. Even more indirectly related is **Chapter 9** in the **Physical** books ("Lithosphere and Tectonics") which focuses on the Andes and other South American features.

Chapter 12 in Blouet "Brazil" can be linked to **Kuby Chapter 6** "Help Wanted: The Changing Geography of Jobs" (activity entitled *Around the World from Pre- to Post-Industrial*). Brazil, as a newly industrializing country (NIC), is one of several used to illustrate differences in economic development levels and can be related to the Blouet activity on the development of Brasilia and the interior since the 1960s.

Chapter 14 in Blouet "Latin America and the World Scene" can be linked to both **Wheeler Chapter 10** "Manufacturing: Where Plants Locate and Why" (activity entitled *Run for the Border: NAFTA*) and to **de Blij Part 11** "Cultures, Landscapes, and Regions of Industry" (activity entitled *Maquiladora? Is That the Latest Latin Dance Craze?*). Both of the ancillary activities deal with Maquiladora plants although the former focuses attention more on the El Paso-Ciudad Juarez area and the latter on the San Diego-Tijuana area.